the layman's weather guide

Pogonips

By

Sheridan D. Jones and Wally E. Kohl

authorHOUSE™

1663 LIBERTY DRIVE, SUITE 200
BLOOMINGTON, INDIANA 47403
(800) 839-8640
WWW.AUTHORHOUSE.COM

First published by AuthorHouse 11/30/05

ISBN: 1-4184-7267-0 (e)
ISBN: 1-4184-4405-7 (sc)

Printed in the United States of America
Bloomington, Indiana

This book is printed on acid-free paper.

DEDICATION

This book is dedicated to two great ladies who put up with two old guys who got the wild idea to write about the weather. Thank you to our wives, Warna and Crichton, for patience and endurance. Warna and Crichton probably would say the best thing about the project is it keeps us out from underfoot. We add a special and sincere thank you again to Crichton for being our first proofreader. Although she caused us many hours of rewriting, we owe much grammatical accuracy to her. With Crichton improvements to our writing we then felt comfortable to enlist the aid of another reader. Carol Possey, an experienced editor with years of newspaper experience, graciously accepted the challenge. Next we thank Stina Seeger-Gibson, an accomplished theater personality and legal secretary, for continuing the proofreading through completion. Their meticulous attention to detail readability has resulted in what is the final product you now have in your hands.

We especially thank the many others who helped editorialize, illustrate, and reconstruct our drafts so they make sense to you.

Many thanks to Ralph Gettleman for endless internet research to locate information verifying the accuracy of our meteorological facts.

This first edition is being released without any of the graphs, photographs, drawings, cartoons, and illustrations contained in our original draft. These will be added in a later edition but are available from the authors by special request.

PROLOGUE

Now that you have our book on weather you need to keep in mind that we are focusing on what happens on the central Oregon coast. There are some useful and interesting facts that are global in nature. Be forewarned however that a direct comparison to an element like wind direction and its implication may not be applicable to other locations. Spend some time in any location and you will soon learn what the wind direction or cloud cover means based on what you read here as provided by Pogonips.

Oh, you are sitting in your chair...it is morning. You say to yourself, "Self...what will the weather be this afternoon?"

Well, you have read Pogonips, so you observe the current conditions, consult your resource charts and predict.

Walla!...;you are right!

TABLE OF CONTENTS

Pogonip is from the Shoshonean. The Shoshonean language is a sub-branch of the Uto-Aztecan linguistic family of North American Indians. It includes Shosone, Commanche, Ute, Paiute, Hopi, etc.

Pogonip means a heavy winter fog containing ice particles occurring in the Sierra Nevada Mountains of the Western United States (CA).

PREFACE

Just so you know what this book is supposed to do for you, here are a few ideas. We are just a couple of old Beach Weather Watchers who live on the North Central Coast of Oregon. Most of the time, it is beautiful, especially if you like rain. So naturally we are interested in the weather and like trying to guess what it will do next. We do the logical things; watching TV, reading the papers, listening to the radio, and reading books; even the telephone book. Believe it or not there are all kinds of tips in the telephone book.

This Oregon coast beach weather may drive tourists nuts, because downtown in protected areas it's sunny and hot; so they decided to sit on the beach and play Californian. Not so fast...when we get on the exposed beach we find the same sun but an icy northern blast requiring all kinds of ingenious work with drift wood and sand mounds to get out of the wind so you can enjoy the sun. We both need more weather knowledge when we head inland. Any direction we travel from here, there could be the risk of snow in the mountain passes, flooding, or wind downed trees; depending on the season.

Even with our collection of texts, some weather definitions are in one book but not in another. This book is a 'layman's' explanation including enough technical information to satisfy your scientific interests. Best of all, you can have fun guessing the weather; occasionally getting it RIGHT. Based on living where these things happen, we will tell you what we think the experts mean and why they say something like "Partly sunny with small craft warnings".

Good luck on weather watching and enjoy yourself using the book. Here is a tip; don't read this book from start to finish. Jump to sections that talk about what is going on when you look outside and something has piqued your interest, then you sit down and enjoy a good read.

1. WHAT IS WEATHER

Daily variations in our atmosphere IS Weather. The general condition of the atmosphere at a particular time and place is weather. Our atmosphere is a mixture of gases in which temperature variations enable the water to be gaseous (suspended water vapor), liquid (rain), or solid (ice/hail/snow).

Weather is the result of variations in humidity, cloud cover, wind, and precipitation. To people, Weather is cold, hot, wet and so on. Weather is miserable, perfect, or tolerable. Why is this so? The Sun! As the earth spins the Sun heats the atmosphere in a highly irregular fashion. The absorption and release of heat energy from the oceans and land masses is responsible for the dynamics of weather. This dynamic, fueled by surface waters, provides the makings for clouds swirling like currents and eddies in a river. Literally unconfined, these clouds can stretch thousands of miles. Now we associate these swirls with high and low pressure movements of air; changing Weather on the ground.

So we have a <u>definition of weather</u>---the state of the air at a specific place and time with respect to temperature, pressure, humidity, cloudiness and 'comfort factors'.

There is high pressure wherever air sinks and low pressure wherever air rises, and winds blow from the high to the low pressure.

Sea breezes cool the air on hot days and warm the air on cold days. These winds are affected by ocean temperatures. What comes across the water is what we feel most. The more wind moves over our skin the cooler we feel. The higher the wind speed the faster the weather changes. This is very important here on the beach.

Trying to write about terms like temperature, barometric pressure, humidity, etc. can be difficult. To really confuse the issue, throw in some terms like 'comfort factor' and 'wind chill'.

Here we go! Remember, weather has three basic elements; the Sun's energy, the atmosphere (pressure gradients), and water. These elements combine to produce clouds, wind, precipitation, and storms.

1.1 Temperature

Temperature is defined as the degree of hotness or coldness of anything, usually as measured by a thermometer. What else could it be? Since that was so simple, let's confuse the issue. Is the temperature in Centigrade, Fahrenheit, Celsius, or Kelvin?

First we have Celsius from Anders Celsius (a Swedish astronomer). In 1742 Celsius introduced a scale with water boiling at 0° and freezing at 100°. In 1745, Carolus Linnaeus (1707-1778) reversed the scale with 0° freezing and 100° boiling water. Centigrade was not named after any person. It is part of the metric system with the prefix 'cent' meaning 100. In case you want a topic for the dinner table conversation, you might want to bring up Baron Kelvin who came up with a scale to measure absolute zero. At absolute zero all molecular activity stops. What this means to us lay persons is, it is mighty cold at those temperatures. Don't worry; you will not be tested on this.

Conversions will be discussed later in this book. For our purposes, just buy a thermometer that has both Fahrenheit and Centigrade. The thermometer will most likely say Centigrade, but to be current with the world, Celsius is the correct term. By now you have figured out these are two names for the same scale.

For our needs we will stick to Fahrenheit, which indicates water freezes at 32° and boils at 212°. By the way, Fahrenheit was named after a scientist Gabrielle Daniel Fahrenheit (1686-1736).

Where does all this temperature come from? The Sun! That old golden orb, which has been worshipped in many ways, is the engine that fuels all life (on the planet earth) and energizes the weather.

In simple terms the sun is a fusion atomic pile that sticks atoms of hydrogen together to make helium. In the process the energy released by the sun is the source of all energy for life on earth and the main engine that powers the weather; both good and bad. Next there is what we call solar flares which meteorologists call solar prominences. These are huge eruptions of solar gas that soar off into space. The solar flares occur in conjunction with those strange things called sunspots. As a result massive storms of cosmic rays and X rays stream toward earth. Even professional meteorologists aren't sure if solar flares affect the weather; let's presume solar flares do affect the weather.

1.2 Air Pressure

Many of us feel sorry for fish when we think about them swimming around in their water world. We know when the fish are at the bottom of some deep body of water, the pressure on them can be tons per square inch.

We too are walking around at the bottom of an ocean. In this case, it is an ocean of air. Despite that fact, we seldom notice the effect of this pressure from the ocean of air we have to wade through but it is quite significant. Think of it this way; for every square inch of surface, the size of a big thumb nail, there are 14.7 pounds of atmospheric pressure.

Next time you're at the gym, pick up a hand weight and load it to 15 pounds and do your exercises. I'll bet it will give you a healthy respect for how many of those 15 pounds are on your body. One reference suggests this amount is a ton of weight on the average person.

1.3 Humidity

Humidity is the amount of water in the air around us. Officially it is the amount of water vapor in the air. This sneaky ingredient serves to confuse our heat sensors. The faster we get rid of water on our skin, the cooler we feel. When the air is warm and moist we feel sticky and say "it is humid." The more water in the air, the less water we can get rid of and the hotter we feel.

Warm air can hold more water vapor than cold air. A cubic yard of air, with a temperature of 68 degrees Fahrenheit, can absorb up to 0.81 cubic inches of water vapor while that same cubic yard of air at a temperature of 50 degrees Fahrenheit can only absorb .44 cubic inches of water. When the air can no longer hold as much water vapor (such as when it is cooled), it condenses out, and this is the dew point. At the temperature where no more water vapor can be absorbed, the air is saturated, and the relative humidity is 100%. Relative humidity is the amount of water vapor actually in the air, compared to the amount it would contain if saturated.

1.4 Clouds

Water vapor condenses into clouds. The coast is a great place to be a cloud watcher. Dark fog and clouds on the horizon, which is over the ocean, could mean rain. The question is how many minutes will it take to come ashore? This will take repeated watching and awareness of the wind, as discussed in the beginning of this chapter.

If you are a detail freak, get a cloud book and impress your friends by talking about lenticular alto cumulus or cumulus congestus. That ought to keep them bored. However, we suggest you begin with the basics, such as the big fluffy ones called cumulus. Then follow that with stringy, layered clouds called Stratus, thin wispy clouds called cirrus, and puffy, heavy, dark clouds called nimbus.

Those are the basic formations. Further study will introduce you to other cloud formations which are combinations of these four basic ones.

Air heated by the ground rises and cools. The result is water vapor which needs microscopic particles in the air to condense around. The result of this condensation is what we call clouds; the big fluffy clouds. Remember cumulus is Latin for 'heap' and there are several varieties; Fair weather cumulus at up to 4,000 feet altitude, swelling cumulus at 15,000-20,000 feet, cumulus congestus stacked from 5,000-30,000 feet.

Cirrus clouds are ice crystals or "mares tails". Cirrus is Latin for curl, hence the phrase describes the fact that these clouds look like streaming curls of hair.

Cirrostratus has a tendency to be way up there; around 20,000 feet. Stratus is Latin for 'layer' or long thin sheet clouds which are thin veils of ice crystals. Cirrostratus is an early warning of a low pressure front called a depression. This is sometimes called a mackerel sky and has the pattern of scales on a fish. "Mackerel skies and Mares tails make ships carry low sails." is one seaman's proverb.

Altocumulus and altostratus are high; ~6,000-8,000 feet. The prefix 'alto-' indicates a cloud which is over a mile high. This cloud form is also thicker than cirrostratus but lower and more cotton ball looking.

Altostratus (gray, mid level clouds hovering at about 12,000 feet) is like cirrostratus but thicker and lower in the sky, so they make the sky look milky.

Cirrocumulus is higher; above 18,500 feet. Stratocumulus clouds (white clouds with dark patches) are lower; up to 6,000 feet. These clouds foretell gradual clearing and fair weather following light rain.

A setting sun reflects off cirrus clouds. Those clouds are probably the tail end of a storm. Red sky in the evening means the view of the sun is through relatively dry air, which can indicate fair weather moving in from the west. If the rising sun shines on cirrus clouds, savvy sailors batten down the hatches.

Anvil cumulonimbus can rise up to the Cirrocumulus height (spiraling to 65,000 feet). The anvil is formed when the cloud can rise no farther, so it flattens as the above wind shears off the moisture. Anvil cumulonimbus and Cirrocumulus appear in combinations and variations. Large blobs hanging down from the cumulonimbus like a giant udder are called Mamma. These indicate strong winds and heavy rain.

Some areas (not the Oregon coast) have lenticular clouds. These are altocumuli lens-shaped clouds around mountain peaks due to a wave in the air stream passing over the peak.

So what do you do with that sighting? Remember! Clouds are fog; fog is water vapor; rain is heavy fog. If you see clouds there is water vapor accumulating in the air. If you have wind, this water vapor may be passing through. You may not get rain, but someone east of you might.

No clouds equal less moisture in the air.

If the clouds are puffy, the air is rising to form these clouds; towering cumulus clouds form thunderheads.

Flat layers of clouds indicate stable air; what you see is what you get.

Big fluffy clouds suggest unstable air and indicate a change in the weather.

High flat clouds indicate stable air, which equals no immediate weather change.

Cauliflower like clouds mean imminent thunderstorm.

Rolling dark cloud indicates the approach of bad weather.

Fleecy white clouds indicate good weather.

High wispy clouds (cirrus) show a front is coming.

There are many cloud variations. Seldom will you see clouds that fit the cloud pictures exactly. Be prepared to generalize when identifying the cloud formations.

There is no accurate way to measure the depth of a cloud formation from the ground that is available to the common weather watcher. Satellites and ceilometers measure cloud elevation and depth.

1.5 Summation

The sky has been a weather forecast source from early times. There is a reference in the bible (Matthew 16:2-3) to "the sky is red" to forecast weather.

The following relates to that Book of the bible; Red sky in the morning, Sailor take Warning (foul weather); Red sky at Night, Sailor's Delight (fair weather). Most of our weather systems on the coast approach from the West so a red sky in the morning means the sun rising in the east is reflecting off cirrus clouds moving in from the west. These thin and wispy clouds are composed mostly of ice crystals. Cirrus clouds typically appear at the beginning or end of a storm and are therefore, early signs of a storm system.

Weather is temperature, air pressure, water vapor (humidity), and clouds. This calls to mind the good stuff like balmy, sunny, breezy, and the like. Weather can also produce the nasty; "I'm freezing." and "Man its hot." phrases.

As has been said before, everybody talks about the weather but there is nothing to do about it. However, it's important to realize weather really is the result of a series of dynamic and ever changing forces that create what happens around us. As long as we stay outside we are affected by weather. The weather is what it is. Depending on where you live, it may be to your liking or it may drive you crazy. The weather's vagaries and whimsical tricks such as blizzards or tornadoes are the same forces that give us clear, pleasant outdoor adventures.

Don't forget, the components are temperature, air pressure, water vapor (humidity), and clouds. These elements, with the sun's influence and earth's tilt relative to the sun's rays, produce the ice, rain, and even the sunshine and balmy breezes we enjoy.

2. WHAT IS CLIMATE

Climate is the overall pattern of all of the meteorological elements; temperature, humidity, wind, cloud coverage, and precipitation. What separates this from weather? Climate is the prevailing or average weather condition of an area over a period of many years.

What is the difference between weather and climate?

Aren't you glad we asked? As best as we can cull from the current lore, weather is what we deal with in the here and now. The word climate is derived from the Greek word Clime. Impressed? Climate means a compilation of weather events over a period of time. This means you can describe your area as a dry climate or a wet climate.

Climate and weather are made up of the same factors; temperature, precipitation, humidity, wind, atmospheric pressure, and clouds. All of these average out to a prevailing pattern that makes up a regional climate, regardless of what it may be doing at any given moment. The norm of a climate does not preclude other weather events from happening.

Don't get defensive when someone says the beach climate is moderate; which to most people means it is nice most of the time. That's true as long as you can acclimate to the long periods of rain-clouded skies. The most important elements forming climate are temperature and rainfall. Obviously these elements don't occur in a vacuum, so it is necessary to interject the factors of topography and latitude. In the case of us lucky beach folks, the ocean is our dominant factor. Sure somebody is going to say the ocean isn't topography. Well, actually it isn't, but the combination of hills, forest, and cities along the shore affect our daily weather and our 'moderate' climate.

Let's not get this too confusing. Look at climate this way. In 1984 we had 10 inches of snow and temperatures of 10°F on the beach. This does not define our climate. Weather can vary widely regardless of what the overall climate may be. Our climate is defined as Coastal. It is temperate which is why most of us like it here. Another way of looking at it is to remember that if you leave your inland house and visit our coastal area, you will find differences in temperature and winds that may be to your liking or in some cases may be affect you negatively. Especially if you forgot that piece of clothing you now need.

OK, one more time. Consider <u>average</u> rainfall, temperature, and cloud cover and you are looking at climate. Climate is the prevailing or average weather conditions of a place over a period of years. A climate zone includes monthly averages and high and low extremes of cloud cover, wind speeds and directions, temperature, humidity, solar radiation, and atmospheric pressures. But climate is more than averages, since it includes extremes and means (the frequency of occurrence).

Just when we decide what climate means to us the weather experts come up with an attempt to more exactly define climate. For example, there is a micro climate limited to a small area; that climate could be the south or north side of your house.

What this means is that there are smaller areas that have a climate of their own. This could be because of some man made structure or a natural object like a sand dune or hill or forest.

2.1. Macro/Micro

World climate is Macro. That is the study of world variations and patterns. These studies involve up to 30 years of data and its associated analyses. Consideration is given to earth's declination, orbit, man's effects, and other broad spectrum concepts.

Micro climate is <u>local variations</u> within the broader area's normal climate. The micro climate may be different from the broader climate because of bodies of water, urbanization (smoke/smog rises to give moisture the particles by which to create rain clouds), vegetation, and topography. If the conditions change, either rapidly or over time, different micro climates may result. This would be a departure from 'normal' climate. In fact, climate is one thing at one altitude and another at a different altitude.

2.2. Patterns

Identification of long term climate cycles is an elusive goal but many weather fluctuations are part of short term patterns. Heat waves, cold snaps, wind storms and other phenomena may be part of a weather cycle and actually a normal pattern of climate for an area. Prevailing westerly winds bring a succession of low pressure systems, associated with strong winds and rain. High pressure systems are more associated with settled weather (calm) and clear skies. So while individual days may have significant fluctuations, we do see developing patterns we can count on. Cycles of three to five years duration have been mapped as variants in ocean temperatures. Links to warm ocean currents (El Niño) become increasingly successful seasonal forecast tools. However, accurately predicting years ahead is still a weatherman's dream.

Other theories for patterns include solar activity. Consider sun spots; the dark area moving across the surface of the sun. The quantities or frequencies of sun spots appear to peak about every eleven years. Magnetic fluctuations on the earth seem to be triggered by the sun spot activity. Lunar tide patterns recur almost every 19 years. Linking weather variations to a combination of these cycles is a science not yet fulfilled.

2.3. Zones

Zones are labels for climatic areas. It is a way to define different areas' climates. If you think we are confused, you are right. The old fashion zones terms like Alpine, Arctic, Desert, etc. are replaced by a variety of listings that attempt to identify what type of climate exists.

The bulb and seed catalogs have a system that is based on when they need to ship you something to plant in your area that will not freeze and yet will bloom that year. These catalogs are big on growing seasons. Sunset Western Garden Book shows Lincoln City influenced by the ocean and the variability of isolated spots. Sunset Western Garden Book text points out that a freeze (which may happen) causes considerable damage to plants if it is early or late. They shows a 20 year minimum temperature range from 28 degrees to 1 degree Fahrenheit.

We are fortunate here in Oregon to have examples of most of the generally accepted climatic zones including Coastal, Temperate, Northern Temperate, Semi Arid, and Arid.

You may have to do some traveling to experience each one. If you drive from the coast to Eastern Oregon you may experience examples of each one. The same is true sometimes just traveling north or south, but keep in mind zones are mostly altitude related. Therefore, another way to experience the zones, to a limited extent, is to start at the base of one of our mountains and drive or hike up. You will pass through a variety of zones all in one locale. You must also understand, climatic anomalies are frequent and you will find variations exist within zones.

Although our area is not accustomed to high winds, they can reach over 100 miles per hour. This is less dramatic when you look at the wind zones across our country.

Coastal = narrower temperature range and weather dependent upon sea surface temperature. Ocean responds more slowly to temperature changes than does land. Sea breezes cool the air on hot days and warm the air on cold days. Wind borne abrasive sand added to the moist salt content of our air creates a harsh environment on the coast.

Temperate = fairly uniform rain patterns and four distinct seasons.

Northern Temperate = similar to Temperate, but the winters are longer with greater snowfall. Mountains may have lower temperatures than low-level locations. Northern Temperate has regular snowfalls and variable rain amounts related to wind speeds and cloud cover.

Arid = consistently low rainfall and very large temperature variations between day and night in summer and winter months.

Semi-arid = higher rainfall than Arid and less marked temperature fluctuations between summer and winter.

Beyond these zones we have Tropical, Subtropical, Mediterranean, and Polar. We shall ignore these other zones, since this is a book about our coastal weather and climate.

For the first time (as far as we know) Oregon has been divided into climate zones.

Zone 1:   The Coastal Area
Zone 2:   The Willamette Valley
Zone 3:   Southwestern Interior
Zone 4:   Northern Cascades
Zone 5:   The High Plateau
Zone 6:   The North Central Area
Zone 7:   The South Central Area
Zone 8:   The Northeast Area
Zone 9:   The Southeast Area

The Oregon state climatologist, George Taylor, has provided these zone designations. Based on his studies, we are The Coastal Area. This he defines as an area that has wet winters, relatively dry summers, and few temperature extremes. His study identifies the nine zones for Oregon, but all we are talking about is Zone 1, The Coastal Area.

For our needs, we have spent our time reviewing the data for the central coast. Buy George Taylor's book to checkout the other eight zones to see if they match your idea of what the rest of the state is like. Here is a teaser about the tidbits you will find in the book. One of the rainiest regions in the world lies between the Santiam Pass and Sisters. In just 20 miles there is a rain difference of 60 inches a year. The next time you make a trip to Central Oregon from the temperate coast, you will be more willing to live with our rainy season and glorify those perfect days that believe it or not, do occur rather frequently here on the Central Coast.

So what does all this mean to us? It can be windy and cool on the beach - most of the time. Mornings can be foggy and clouds may form on the horizon at sunset -most of the time. Dew can be found on automobiles parked outside any morning -most of the time. By the way, that dew is laden with corrosive salt; wash it off daily. Wind comes from the south in winter, perhaps with a storm attached and from the cooler north in summer - most of the time.

2.4. Our Backyard

There is, however, another term in use now, Bioclimatology. Bioclimatology includes reactions and behaviors of plants and animals. Factors that make up a micro climate, such as heat, humidity, atmospheric pressure, sun light, winds, and precipitation need to be addressed.

One of the goofy things we do when we get to the beach from some other climes is to start gardens expecting our favorite plants to grow here. Many of them do grow and flourish, like Scotch Broom, the hardy bush covered in bright yellow, fragrant blossoms in spring. This is an example of a plant introduced to the coastal area that stayed and thrived. It has become a coastal pest and was blamed for the famous Bandon, Oregon fire. This town was surrounded by the highly flammable scotch broom and was mostly destroyed by fire.

Ivy likes it here on the coastal Oregon and before you know it your trees are being strangled to death, or you can't find the garden path. Both the solitary and the creeping types of Nasturtium go nuts around the beach, especially if there is no hard freeze. Don't forget, you can get even with them by eating the Nasturtium flowers in your salad.

Plant your garden with some native species based on the climate and the micro climates at the coast; Salal, red current etc. It makes more sense. Some of them, like honeysuckle will take over, so be careful where you plant. Your local nursery will help you select colorful plants that grow successfully.

Another thing to keep in mind is that you are providing food for the beasts that range in size from your friendly garden slug to the local deer herd. These animals feed on the vegetation in your backyard. Roses are a favorite for deer and there are all manner of old wives tales about what they will not eat and what will keep them out of your yard. Don't bet on anything working; share your wealth with the animals by planting early and enough, so that the animals can not eat it all.

So based on the principles of Bioclimatology, your plants will need to be selected carefully, taking into account the overall climate and exactly where you wish to plant (the micro climate).

Take a walk around the outside of your house or from one window to the other, including any balconies. Whether or not you planted anything in your yard you will find some things grow on one side of the house as opposed to another side. During a cold snap take another tour and you will find some plants are zapped dead, while the same plant in some other spot in your yard is perfectly OK.

Cold air is heavy and moves down hill so, if your house backs up to a rise, that side of the house will have a cold spot. Obviously the side of your house with the sunshine will have a warmer climate that, in the summer, may kill your plants. The other side of your house may not even have enough sun to make plants grow. Also watch your animal friends like our local cats. On a cold day they will rest next to a sunny outer wall out of the wind and be perfectly happy. This will give you a clue as to a perfect planting spot.

Let us not forget the wind. It prevails from the north in the spring and summer and from the south in the winter. This will, among other things, affect how fast plants dry out.

Be aware that there is a microclimate in your specific location that will guide you in planning the orientation of a new home or deciding what to plant in which location.

2.5. Summation

Climate is all the elements of temperature, humidity, wind, cloud coverage, and precipitation that we described as weather. When these measurements are taken over many years to establish averages for the seasons, we call that climate. For our North Central Coast region of Oregon is a temperate climate; one of few extremes.

Macro refers to the world climate, while Micro refers to variations by locality. These smaller geographic areas of climate are sub-climates; thus the term Micro. The macro aspect of weather or the world variations and patterns come to us via our favorite TV weather guys.

Long term cycles define "patterns" and zones as the areas having nearly identical climates. The patterns we have on the Central Oregon coast will begin to identify themselves. Most frustrating is the fact that conditions in these zones don't hesitate to change often.

All of these phenomena work together, to make our central coast climate fit the definition of temperate coastal. Analyze your location before you plant your favorite flowers. All in all, after a while you will begin to identify for yourself the aspects of weather that are characteristic of our Central Oregon coast climate. We will bet most of you (once you get through the dreary drizzles) say, "Why didn't we come here sooner?"

3. ATMOSPHERE

That big ocean of air that we wade around in all our lives (mostly at the bottom of) is called the atmosphere. The atmosphere is said to be the creation of cosmic dusts of predominantly hydrogen and helium. It is composed of 21-23% oxygen, 76-78% nitrogen, carbon dioxide, a lot of other gases (Argon is the next most prevalent), and some water vapor. Earth's gravity does little to pull the hydrogen and helium out of the atmosphere. Add to this the byproducts of earth's surface activity from volcanoes to smog and you get nitrogen and carbon dioxide. The volcanic action starting below the surface of the earth, plus organic decay adds the compounds of methane and ammonia. Plant life produces oxygen from the carbon dioxide.

The atmosphere is composed of layers. The layers of our atmosphere have names that most of us can't recall, like exosphere. These layers, starting at the surface of the earth, up to the edge of space are defined in the following sections.

3.1 Troposphere.

The Troposphere is the relatively shallow layer we live in and is best described as the weather zone. It is the most changeable layer since it is the only one containing enough water vapor to form clouds and is located from zero up to seven to ten miles above the earth. Most of the clouds and weather are formed in the Troposphere. The temperature drops about three and one half °F for every 1,000 feet of altitude. Believe it or not, that temperature drop is about the same whether you are in a desert or on the ice cap. In the Troposphere high level clouds (averaging 20,000 feet up) are called Cirro-form. These are the delicate, wispy accumulations of ice crystals.

Tropopause is the boundary between two layers, and is not the same height in all parts of the world. For instance the warm air of the equatorial area may push the Tropopause up to 10 miles above earth. Over the Poles it drops to about five and a half miles. It isn't a smooth layer but a series of plates with very strong, extremely fast winds blowing along the breaks. These winds are known as the Jet Streams. They are "tubes of air movement" thousands of miles long and hundreds of miles wide with a core speed of 85 mph or more. Jet Streams are normally westerly (from west to east) direction.

3.2 Stratosphere

The stratosphere is 7-30 miles high with relatively little water vapor or dust since this layer does little mixing with our Troposphere. It is less turbulent and without water vapor to form rain, hail, snow, or dew. In fact, there is no weather at all in the stratosphere; no clouds nor up and down air currents. The temperature averages -70 °F until you reach the next level, then a strange thing happens. Ultraviolet sun rays begin to be absorbed and guess what? The temperature soars up to 170°F+. This is called the Ozone layer or the Ozonosphere.

3.3 Chemosphere

Chemosphere is the name given to the next area and ranges from 20-50 miles up. It is mostly water (H_2O) but short wave radiation from the sun breaks H_2O and CO_2 molecules into hydroxyl (OH) - not a very habitable area.

About 40 miles up, the Ozone layer absorbs the short (ultraviolet) wavelength radiation, which is harmful radiation energy given off by the sun. Thus the Ozonosphere protects us. This is a thin layer of the atmosphere. Above the Ozone layer the temperature begins its steady drop, but an inversion is coming in the next layer. Make sure your travel agent knows this if you are planning a commercial space trip, commercial trips are being planned.

3.4 Mesosphere

The Mesosphere is 30-50 miles up with a temperature averaging -90° F. No current aircraft can travel at that altitude.

3.5 Thermosphere

The Thermosphere is about 50 miles up. Temperatures there range from 960°F to more than 3,000°F. It depends on the solar action since various molecules absorb solar radiation. This is rocket and satellite country.

3.6 Ionosphere

On up to the Ionosphere (60-100 miles up), the atmospheric pressure is extremely low. In this layer molecules of various gases begin to get bombarded by short-wave radiation which causes the atoms to lose electrons. This is ionization. This causes radio interference and may be the origin of the aurora borealis. Some books say this layer is part of the Thermosphere.

3.7 Exosphere

The outermost shell of our atmosphere is the exosphere and this ends what we call atmosphere. No one can say reliably how far up it extends. It could be 500 or more miles, since it just gets thinner. There probably isn't a specific end as it blends into deep space. Temperatures are thought to reach 4000° F. at some 400 miles above earth.

3.8 Summation

That is enough of that 'gobbledygook', but we had to tell you so you know your weather world is part of a complex system. Scientists speculate our entire atmosphere originated as cosmic dust; a blend of hydrogen and helium. Earth's chemical reactions produced nitrogen and carbon dioxide, and earth's gravity, spinning, and surface irregularities resulted in the layering we have defined.

OK, you've noticed the boundaries very between the layers. Suffice to say it is not exact, nor does it remain constant over time. It's merely an attempt to classify layers to better understand the dynamics brought into play which affect the weather. The importance of all the layers of atmosphere is how the sun's rays affect what happens to the weather at ground zero.

The sun is the maker of weather on earth. The atmosphere is that envelope of air we live in and enjoy. We enjoy it, that is, as long as it doesn't play any of its violent tricks on us. Lucky for us it rotates with the earth, because if it didn't, we would always be facing a headwind. Especially since, at the equator, the earth's rotational speed is ~1,040 mph. To be more precise, the atmosphere is free to shift as the earth moves and trade winds result. These winds rotate in circular patterns (clockwise in the Northern latitudes and counter-clockwise in Southern latitudes); because of the Coriolis force.

4. PRECIPITATION

Is it raining, snowing, hailing, feeling like a wet cloud around us, or is there dew on surfaces? The form precipitation takes is dependent upon whether the air is cooled below its saturation point. Colder air is dryer. We will talk more about this later. We are talking about moisture in the air that falls to earth.

4.1 What It Is

Precipitation is classified according to the form it's in when it reaches the ground. Precipitation is rain, snow, sleet, or hail which can only occur if there are clouds in the sky. We need the presence of tiny foreign particles around which the moisture can condense. These particles may be dirt or even tiny ice crystals and are the necessary nuclei for droplets to form. The process of raindrop growth is referred to as coalescence; the act of uniting and growing together.

Temperature will determine whether precipitation will occur. The type of precipitation that reaches the ground depends on the process that happens within the cloud. Air is heated by the ground and rises and cools which causes water vapor to condense around those microscopic particles. The temperature of the air between the cloud and the ground is the deciding factor. Shorter clouds in warmer air yield drizzle and the billowing clouds allow the drops to get bigger.

Remember, all clouds are rain and all rain is also snow.

The droplets in a cloud are about 1/100th the size of a raindrop. They freeze into ice crystals or snow. The density of freshly fallen snow is about 1/10th that of liquid water. The only way each droplet or crystal can fall to earth is when it gains enough size and weight.

When heavy enough the droplet's weight overcomes the lift of moving air and down it comes. Gravity again! So the droplets are tiny ice crystals (snow) or water unfrozen. If the air is colder at or near the surface, we get freezing rain. If the clouds are even higher, so the moisture particles have farther to fall, the air temperature determines if dry (very cold) snow or wet snow falls. The droplets start as ice or snow and turn to rain as they fall. Warmer temperatures melt the droplets into raindrops. Raindrops act like prisms and result in our rainbows.

4.2 Classes of Precipitation

Precipitation is classified as steady or intermittent. Clouds generating rain are commonly in the nimbus (rain) category. OK, let's match the rain to the type of cloud it comes from.

Heavy rain   ~ cumulonimbus
Steady rain   ~ nimbostratus

Widespread stratiform clouds such as Stratus or altostratus generally produce steady rain or snow.

Light rain; cumulus congestus

Showers (sporadic rain); mostly from cumuli form. Cumuli form clouds <u>tend</u> to produce intermittent precipitation.

Is it Rain Drizzle or Mist? All these are a measure of quantity and intensity so we create a table from cloudburst to fog.

- <u>Cloudburst</u> is big drops about 2.85 millimeters (mm) average and up to 6.35 mm which accumulate up to 4 inches an hour.
- <u>Excessive</u> is slightly smaller drops (2.4 mm) at about 1.6 inches per hour of accumulation.
- <u>Heavy</u> is 2 mm drops making about 0.6 inches an hour. Heavy rain seems to fall in sheets, thus reducing visibility.

- Moderate is 1.6 mm diameter drops but only 0.15 inches per hour
- Light rain has 1.24 mm drops and about 0.04 inches per hour
- Drizzle is less than one millimeter in diameter ;<0.02 in. and falls at approximately 0.01 inches per hour; so it appears to float in the air. Drizzle is light, moderate, or heavy based on visibility.

  If visibility is less than 5/16 mile, it is heavy drizzle.

  If visibility is more than 5/8 mile it is light drizzle

  Drizzle is fine drops falling close together.
- Mist is 0.1 mm drops and almost no measurable accumulation. Mist is **not** falling so is actually a light fog. Visibility is restricted, but if visibility is a half-mile or more, the fog is known as Mist.
- Fog is clouds at ground level.

Since the moisture content is closer to vapor than to rain drops, accumulation of moisture is imperceptible, so your rain gauge on your weather station will not register. Fog is condensation that occurs when the air can hold no more moisture (saturation point). Drifting or blown moist air may result in fog over water or on hillsides such as in the mountains. Most fog is the result of heat radiated back from the earth producing condensation (radiation fog). Radiation fog is always found at ground level but its depth may extend to about 1,000 feet above ground level. If a thick layer of moist air is present we have fog. Add smoke to fog and, yes, we have Smog.

If only a thin layer of moist air is present it is Dew. However, it's not possible to have fog without dew. Dew is not really precipitation but since we think of it as such it is included here.

4.3 How Precipitation Is Formed

Not all kinds of clouds can produce precipitation. The water vapor molecules need something to stick to when they form. The "something" that enables water to condense into droplets in the air is called "condensation nuclei". These tiny particles are matter such as sea salt in coastal areas like ours. The salt particles actually attract water. Salt from the sea is an important condensation nucleus as far as 100 miles inland. Salt shoots into the air when bubbles burst in the ocean and the wind picks up salt as it blows foam off breaking waves. How small are these particles? The size is about 0.1 micron in diameter, up to a maximum around 10 microns. So how big is that? Well, a human hair is 100 microns thick.

Relatively large salt particles explain the haze along the beach as water collects on the salt nuclei. The water drops collide but do not stick together. Thank heaven it isn't smog. We all know fog when we see it but beach fog was a mystery to us. It can occur along the beach, yet a block inland it may be clear. This is truly a coastal effect. Lucky for us it's not long-lived or very dense. The salt haze causes a somewhat unique, worrisome condition which has some of us driving rusted out vehicles. Here's a practical hint: wash your car often to keep most of the salt off of it. This will reduce chances of coast metal cancer.

Other condensation nuclei come from sand and dust blown into the air, particles in smoke from fires, material ejected by volcanoes and even tiny meteorites that burn as they enter the atmosphere. Material put into the air by people - pollution - can add condensation nuclei to the atmosphere. This can make polluted air hazy, but haze doesn't mean pollution at the coast.

Haze in clean air is likely when humidity is close to 100%. In clean air, condensation does not take place easily.

The process still requires those particles called condensation nuclei.

Precipitation that falls from clouds is made up of larger water drops, ice crystals that have melted, or still frozen crystals. Sometimes all three fall together. Gravity pulls the cloud droplets toward the earth.

Pollution is with us no matter what. Coastal wise, since there is little manufacturing, fewer cars and the demise of the wigwam burner (a wigwam shaped structure that burns waste wood at a lumber mill), we are not affected as much as the cities or valley areas. Once in a while a slash burn will cause a smoky haze that permeates our pristine air. Occasionally a controlled demolition burn will put particulate in the air. The other side of the coin is that without particles in the atmosphere we would not get those beautiful and colorful sunsets and the occasional 'green flash.'

4.4 Humidity and Dew Point

Humidity is the term used when talking about the amount of water vapor in the air. Humidity is unseen, in contrast to precipitation which is water (not vapor) in the air and usually visible in the form of rain, snow, etc. Precipitation is moisture falling rather than floating in the air.

About 90% of the water vapor in the air comes from the oceans through evaporation. The warmer the air the more moisture it can hold until it becomes saturated. If the air is moist, it suggests rain, and it should feel warmer, but it could also suggest snow. If the air is dry it will feel colder because the moisture on your skin evaporates and lowers body temperature.

Unsaturated air continues to absorb water vapor but when the air reaches the point of no longer being able to absorb water vapor, it has reached it's 'saturation point'. Warm air can hold more water vapor than cold air.

Absolute humidity is a measure of the volume of water in a specific amount of air at the current temperature. Absolute humidity changes exponentially with temperature, so we use the phrase, relative humidity. The amount of water vapor in the air (at the current temperature) as a percentage of the amount to reach saturation is Relative Humidity. Saturated = 100% so 75% is three-quarters of its capacity.

No matter how you define humidity, it is the effect it has on us that counts. Our bodies only react to differences in temperature, wind speed and the resultant changes. So, when its cold and dry, we react because our lips crack, our nasal passages feel dry, etc. When it is warm and moist we start using words like 'muggy' or "Man I'm all tuckered out and its only 10 a.m." Once again, the proximity to the ocean means a more or less constant supply of moisture laden air. We don't rely totally on the rain for our humidity, however, when it rains, the moisture content goes up and so does the humidity. Rain is drizzle, mist, snow, hail; its all falling moisture but what if it is not falling or condensing out of the air? What if the air is not saturated? We still need to know the water vapor condition of our atmosphere.

As we have explained, air can hold only a given quantity of water vapor. When it reaches its saturation point the water will then condense (form a liquid). The temperature at which this condensation occurs is known as the 'dew point'. Cool the air and it has to give up some moisture and that is its 'dew point.'

Obviously the temperature has to be freezing to turn the pearly drops of dew into Jack Frost's handy work. Along the coast, when frost occurs, enjoy it. Maybe, just maybe, there will be a cold snap that will let old Jack Frost paint the inside of your windows. Maybe he will be a playful and paint the entire car with frost. It is also obvious that frost can be very specific as to where it forms. For example it may be on one side of your house and not the other or only on the roof tops in your neighborhood.

Most of the time you only are able to enjoy it on the ground and tree leaves as it sparkles in the sun before fading away. The weather is likely to be fair for the next 12 hours if there is a frost or dew formed early in the morning or late in the prior evening.

In cold climates dew point is usually in the low 20s but it is of no consequence because the outside low temperature gets your attention first. It is also obvious that frost can be very specific as to where it forms. For example it may be on one side of your house instead of the other or only on the roof tops in your neighborhood.

Snow will form if the air is cooled below its dew point without water vapor condensing. The temperature has to be below the freezing point of water (32° F). When a thin layer of moist air near the ground cools below freezing and forms ice crystals, we have frost (known as true frost or hoar frost). This is a sure sign it's getting colder and plants as well as insects will be killed. Remember the frost is part of the natural cycle. We need it to keep insect populations in balance; despite the fact that it kills some of our less hardy plants. Plants may do better next spring because the frost cuts down on insect damage and fungicide pests.

While gathering information for this section of the book we discovered our analog systems do not give dew point readings. Digital weather systems seem to always provide dew point. Dew point temperature is the temperature at which the moisture in the air condenses and forms dew or frost (more correctly hoar frost). If this condensation occurs at the ground it is Dew which is water vapor condensing on the surface of solid objects such as the ground. Ideally this indicates higher humidity near the ground, which should produce clear air day or night. When the dew point temperature and air temperature are equal, the air is said to be saturated. Dew point temperature is NEVER GREATER than the air temperature. Therefore, if the air cools, moisture must be removed from the air and this is accomplished through condensation. This process results in the formation of tiny water droplets that can lead to the development of fog, frost, clouds, or even precipitation. (http://ww2010. atmos.uiuc.edu/(Gh)/guides/maps/sfcobs/dwp.rxml).

The best source we have found for dew point temperature is going to this on-line source, http://www. decatur.de/javascript/dew/. Using the temperature and relative humidity, it will compute dew point.

The obvious question was, what does knowing the dew point temperature tell us. Raise the temperature and the air will hold more water. If we lower the temperature below the saturation point the air will give-up some of its moisture in the form of dew or precipitation. If the dew point temperature and the outside temperature are the same, it means 100% humidity. This suggests that clouds or fog will form. We always wondered how the weather guys guessed when we would get fog.

Next, our research pointed out that comfort factor and heat stress indices are influenced by a combination of temperature, humidity, and dew point. For example, when the dew point reaches 60°F most people are uncomfortable. If the dew point temperature reaches 70°F most people would call it steamy hot or humid. One text we read stated, if you have heart or respiratory problems you should steer clear of climates with dew points in the 70°F range most of the time. An 80°F dew point puts most people out of action. If you are forced to work outside, be sure to drink plenty of water or better yet, find a shade tree, and relax until the condition moderates. A dew point of 50°F or lower is generally pleasant for most people unless the outside temperatures are too high or low.

4.5 Ice - Hail - Sleet (Graupel) and How Wet Is It

Now lets move on to solid forms of precipitation. These are more observable and the variety is quite remarkable.

There are 10 kinds of <u>frozen</u> precipitation. "Snow" types account for seven of them. These are part of an international snow classification system. The other three are Graupel, Ice pellets, and Hail. The common word "sleet" refers to frozen raindrops or ice pellets but it is **not** an official term.

Even in areas where snow is rare you can still see it in the sky. Cirrus clouds are ice crystals. However their precipitation never reaches the ground because it melts as it enters the lower, warmer air. Precipitation that evaporates before reaching the ground is called 'Virga.' Hailstones, though, may reach the ground even at warm temperatures.

In any case we are talking now about the kinds of ice that fall from the sky. Snow is crystalline and takes many forms.

Speaking of forms of snow, we asked the curator of the Arctic Museum in Monmouth, Oregon for some help. She sent us lists of snow names and ice names used by the Eskimos. Since snow and ice are so important to their world they have to categorize the snow and ice to convey useful, if not critical, information. Approximately 20 names are used for each. For example, snow can be fresh powdered, pack, rippled surface and so on. Ice is labeled thick enough to walk on, in the process of breaking up, as well as simply thick or thin.

You may have seen grainy snow or soft hailstones. That is graupel; a mass of frozen cloud droplets. Sometimes the droplets are formed from a cluster of ice needles which gives graupel a conical shape. It is soft and may be lumpy or somewhat spherical, but upon landing it flattens into a rounded spot of powdery snow with no apparent crystalline structure. Graupel is often found in severe lightning storms since the particles may be highly electrified and these 1 to 67 mm particles are frequently the core for hailstones.

Hail begins as a frozen rain drop that is kept from falling by the updraft in a thunder storm. However the updraft merely exposes the 'rain drop' to super-cooled clouds and hailstones are formed. Thus, the hailstone forms in concurrent layers as it falls, and then rises again. Weaken the updraft or just let the drop get big enough and down it comes. The smallest hailstone is about two tenths of an inch in diameter, and the largest hailstones are larger than a softball.

Here are some standardized definitions:
Dime sized = 0.50 inch
Penny sized = 0.75 inch
Nickel sized = 0.88 inch
Quarter sized = 1.00 inch
Half Dollar sized = 1.25 inch

4.6 Fog

Where would we be without fog? No spooky Halloween or Jack the Ripper movies, fewer deadly respiratory problems in cities, fewer multiple-car crashes on the freeway, and pilots not disoriented by lack of visibility. However, here on the beach it's a nice change to have a quiet, cool fog roll in and help you feel secluded as you walk the beaches. Our fishing friends these days with LORAN and GPS (direction finding systems) don't fear fog as much as they used to.

We are most likely to get fog on the beach when the inland valley is hot and we don't have a wind to blow it away. Most people who come to the beach to get away from valley heat waves say it's such a relief to motor down the west slope of the coast range into fog.

What is fog? As usual we will give you the simple answer now and try to confuse you later. It is water vapor (clouds) on the ground. Like any cloud, fog is water vapor that has condensed into droplets big enough to see, and there are ice fogs which mean the droplets have frozen. Remember the name of this book is Pogonips. Pogonips means a heavy winter fog containing ice particles, often occurring in the Sierra Nevada Mountains of the Western United States. That type of fog is not likely to happen here on the central coast.

Fog is classified according to its method of forming; radiation or advection. Radiation fog (also known as ground fog) happens during cloudless nights when moisture laden air settles into the valleys. The earth's surface cools rapidly and the air temperature reaches the dew point. It usually dissipates shortly after sunrise even around Cascade Head just north of Lincoln City. Cascade Head usually banks up the fog and causes it to move out when a north wind blows. Radiation fog is the more localized of the two.

Advection fog is caused by warm moist air traveling over colder surfaces, such as the ocean waters, and gradually cooling as it moves over higher terrain. This fog can be dangerous and disruptive as surface visibility is reduced to near zero.

In both radiation and advection fog, the principle is cooling air to its saturation point. Evaporation fog is formed when excess moisture is pumped into the air and overloads the air's capacity to hold water.

That is enough on dynamics. Enjoy this beach weather phenomenon but, be safe, keep track of your location.

4.7 Summation

After all that detail the only thing to say is rain comes in many sizes and in different numbers of drops at any given time. Most Oregonians have learned the word drizzle; which means enough moisture to use the intermittent wipers, but not enough to put up the umbrella or put on a rain coat. You walk a little faster into the office or store. At least that is what the locals do. More than that, add wind and you have horizontal rain or "who pulled the plug?" Other parts of the country use words like 'gully washer', 'cats and dogs', or 'the stuff really hit the fan'.

The highest annual rainfall recorded is 467.5 inches in Mawsynram, Meghalaya, India. The greatest 24 hour rainfall was 73.5 inches March 15, 1952, in Chilaos La Reunion, Indian Ocean. The largest hailstone ever measured was 2.25 pounds April 14, 1986, in Bangladesh. I'll bet that would put a dent in your Volkswagen. The heaviest snowfall was 451 inches in March 1911, at Tamarack, California while the greatest measured snowfall was 1,224.5 inches February 19, 1971, to February 8 1972, at Paradise, Mount Rainier, Washington.

A genuine native Oregonian still won't use an umbrella. However, they do hustle from their cars to lessen the time of exposure. Our experience is that you might as well stroll along because you'll get wet either way. Another problem with an umbrella is when it rains around here the winds may tear your umbrella inside out. Another thing to notice is that no one uses galoshes, and very few use rubbers over their shoes; we just splash merrily along with wet feet; i.e. web foot.

5. COMFORT FACTORS

Journalistic meteorologists have coined a newer phrase; Comfort Factor.

This is to help us estimate how we will feel when going outside to face the daily commute or outside job; or, better yet, a fun, outdoor recreational activity. The combination of humidity, temperature, and wind, all combined make us feel generally comfortable or uncomfortable. Comfort factor is purely a subjective guess by our weather people to help us prepare for the day.

If you live in the Arctic and the summer temperatures reach 50°F you'll feel hot and perhaps uncomfortable. If you live in the Caribbean and casually mention to a tourist that it is a great day at 85°F and blowing wind, the tourist will think you are mad with heat stroke. It really amounts to what you are used to.

There is a weather range, regardless of where you live that lets you feel comfortable. Texts suggest an ideal combination as 75°F and 55-60% humidity with light air movement of 4.4 miles per hour.

Comfort is like beauty; it's based on the 'body' of the beholder. What is comfortable for one person is not necessarily so nice for another. There appears to be no chart using the word 'comfort factor' so it remains a concept. You know what you mean when you talk about it.

For our needs we will accept the idea that the comfort factor can be high or low based on how the various elements of our weather come together at a given point in our lives.

5.1. The physiological mechanism

Everybody knows about calories and the body's fat layer. It is clear when we eat, our bodies have energy to get us through the day and keep us warm. Scientists and fitness trainers use the words "burn fat for energy". Don't pin them down about what is meant by 'burning.' All we know is when you work hard or exercise, you "burn fat for energy". This keeps us fit and trim. Heat comes from our internal furnace fueled by food and moderated by the various ways the body has to radiate heat.

On the cellular level it is easy. Food enters the cell through osmosis and is burned. In the process heat is generated which warms the body. If you are one of those persistent people, the next question is how does it burn and produce energy. Literally, there is an atomic reaction (cell fission) where atoms exchange electrons in their orbits and in the process release energy. Now you won't laugh the next time you read in 'Believe It Or Not' that someone spontaneously combusted into a pile of ashes. Whether it is true or not, the fact remains, in the atom level of the food burning process, it is pure fission or splitting of the atoms. That is the source of the heat in our bodies.

Now the next question is how we regulate that heat so precisely. The best place to start is with the body's thermostat. It's called the hypothalamus and is located deep in the brain, on the bottom, above the pituitary gland.

One group of nerves in the hypothalamus is concerned with the control of body temperature. The nerves are sensitive to the temperature of the blood flowing through the brain, so the brain can switch temperature control mechanisms on or off. What happens next is a series of things most of us are well aware of if we get hot, or too cold.

Let's start with 'too hot.' Lucky for us we aren't dogs. They can't get rid of moisture through their skin so they use their tongues. We would look pretty silly panting along with our tongues hanging out, so we sweat. Even women do it. Perspiration evaporates and when it does it helps us cool off. It lets the heat out. By the way, when blood in our capillaries rises to the surface to aid in heat dissipation, this causes us to appear flushed. That's about all we can do about being too hot, except to head for cooler environs; or take off clothes.

Too cold? Several things will happen. One is the shiver reflex. If we get too cold, but not dangerously so, we shiver, this stimulates blood flow which warms us. Goose bumps pop up but are a lost cause. The hair follicles would rise up to fluff up our coats to trap warm air next to the skin. Since we do not have that thick a coat of fur anymore, it doesn't do any good. The old reflex still does its thing so we have goose bumps.

Eskimos are ideally built for cold climates with short torsos and short arms and legs. That keeps everything close in, to stay warm, since the blood doesn't have as far to travel. The natives in some areas are long, rangy, and thin which allows them to dissipate heat as there is more surface to cool whenever a breeze comes along. Skin color comes into play in a less significant way to maintain temperature. We know Vitamin D is absorbed from the sun, but this can be overdone. Individuals in areas such as Scandinavian countries are fair skinned to soak up more Vitamin D, while people in sunnier climates tend to be darker skinned to reduce Vitamin D absorption.

The extremities usually get the coldest. Unfortunately for us, if the hypothalamus senses we are pumping cold blood to the capillaries in the extremities, it closes that part down keeping the blood closer to the body core. In extremes like drowning in a frozen lake the body shuts off blood supply to other parts and keeps warm blood for the brain stem. So called 'dead' people have been revived because the brain saved itself.

The one thing left to say is what a marvelous machine the body is with its intricate mechanism, that given half a chance, will maintain a constant 98.6°F regardless of outside temperature.

5.2 It Is So Nice Out; Why Am I So Cold?

Most of us have taken that picnic swim at the lake. If you have been sunning yourself and you jump into the lake, it feels icy cold. If it's breezy and cloudy the water seems warmer than standing on the beach. This happens even though your brain tells you the lake is still the same temperature. Heaven forbid we fall into the ocean which is usually (winter and summer) around 50°F. We could suffer hypothermia in short order since the ocean (or any body of water) removes heat from our bodies at an alarming speed.

The heat sensitive regulator for our body temperature is the Hypothalamus region of our brain. However, the sensors play tricks on us because they do not tell what the temperature is; they tell us the difference in temperatures between the last space we were in and the new one. Our body's heat sensors are accurate in terms of letting us know what we need to know to stay alive, or at least comfortable. When the sun is out we feel better, regardless of the actual temperature. It does not seem to be as chilly, especially if there is NO wind.

Along comes the wind to evaporate that moisture on our skins and now we are faced with the wind-chill factor. The more wind that moves over our skin the cooler we feel. Our bodies perspire as part of the heat regulatory system. The faster we get rid of moisture on our skin (by evaporation), the cooler we feel. The sneaky ingredient to confuse our heat sensors is humidity; the amount of water in the air around us. The more moisture in the air, the less moisture we can get rid of and the hotter we feel. The old wives tale, about dry climates not being as hot as moist climates is true. Since we are close to the ocean there is usually moisture in the air.

5.3 Clothing

Another issue that affects our comfort factor is our choice of clothing. Outdoor clothing magazines say layering is the ideal way to dress. In our case all we can do is add on layers or take off a layer. Dress, using the layered look, to protect yourself from the elements and you can always be comfortable on the beach. We realize this is almost heresy, since most people wear what they think looks good not what they ought to be wearing. Also it flies in the face of customs which dictate neckties etc.

Some cultures wear very little and others cover up completely. Both are accomplishing the same thing; regulating body temperature.

5.4 Wind chill

Wind chill is another general measurement that has been added to the list of information available from newspaper, radio, and TV weather people. It is a measure of how cold it feels. When the wind blows the warm air from around our bodies we naturally feel colder; not to mention the potential of increased evaporation from the skin. When this happens we feel colder than the given temperature. In colder climates it could mean frost bite even though the temperatures are nowhere near freezing.

On the Oregon coast we normally don't have freezing weather, but you can see from a wind-chill chart that on a chilly 30°F beach walk with our ever-present winds, exposed flesh could be in danger. You must pay attention to what is going on around you. You don't have to be in Siberia or the Yukon to be affected by wind-chill.

Don't try to guess the wind chill other than to take precautions to reducing the amount of time you spend exposed on a windy, cold day, and to be prepared to add additional clothing to protect your face and extremities. Your body will react to the wind-chill factor. The problem is your body may not tell you that you are in the danger zone. Notice on a wind-chill chart, that not much wind is needed to turn 30°F into 27°, 16°, or 4°. Here is how they match. Only 5 mph puts it at 27°; double that to 10 mph and you are down to 16°; double it again (20 mph) and it is 4°. That is cold even if the table exaggerates the effect. It is also cold enough for frost bite, which if not cared for quickly and carefully can lead to serious damage.

If possible, request a copy of the latest wind-chill chart from NOAA. See their web site.

5.5 Humidity

The layman's definition of humid is when you walk outside and feel clammy. When your lips feel dry the humidity is down. It is also said some people can smell a storm coming. We smell the rain here on the coast, but that this is not very accurate since the humidity is up most of the time anyway.

When someone lays one on you about "It's the humidity and not the temperature." They are correct. The amount of water in the air is humidity. The more water in the air, the less moisture your skin can get rid of, so you feel warmer. In a dry desert climate where you sweat, and if you don't mind the wet feeling, you don't feel as hot.

If you're trying to save money on utility bills, be sure to keep the humidity up inside your house. An open container of water will help. However, that is a losing battle if you have a furnace with a circulating fan, since it removes moisture and defeats the natural effect. However, we are living in a humid environment here at the beach, so dehumidifiers help us attain that 55-60% 'ideal humidity'.

5.6 Air Pressure

Barometric pressure is the weight of air on the body. Therefore certain parts of the body, like an arthritic joint, may react to a pressure change. Despite aches and pains or whatever, changes in barometric pressure are not normally detectable by our usual senses. Most of us can't detect even the normal high and low pressure that moves the air around us. The air weighs you down at sea level, at 14.7 pounds per square inch.

So, get yourself the measuring gear (an aneroid barometer) and enjoy the highs and lows as they happen on regularly irregular cycles.

5.7 Summation of Comfort Factors

The comfort factors of relative humidity and temperature are probably the easiest to measure.

As part of the attempt to quantify comfort factors, there are charts that allow you to prove to yourself that you feel comfortable or otherwise. Look at Wind Chill, Heat Stress, and Heat Stress Index charts in the weather books listed in the bibliography. The heat stress chart is based on the relative humidity and air temperature. The heat stress chart is also based on the fact that regardless of whether you are inside or outside, you are not in direct sun light, and there is a very light breeze. The Danger category should not be taken lightly. Any extremes in temperature that our bodies can not easily compensate for can be deadly. The air weighs you down at sea level, at 14.7 pounds per square inch. If you combine the wind chill chart and the Heat stress chart you come close to having a means to measure comfort factor.

6. PROFESSIONAL METEOROLOGISTS

There are many tools for the modern weather person. The computers correlate, translate, collate, relate and on and on. A person (or a committee) still must decide what to say with all this data. Remember the best weather forecast results in this scenario.

You are on one very high hill (mountain) with your cell phone. The wind is at your back; you are facing the hill (mountain) across the valley where your friend lives. The rain hits and you are getting drenched but your friend's hill has sun. You call him and say, "Hi friend, thought you should know it is going to rain on you very soon!"

This simple example is needed to remind us of the importance of the human element. Many sophisticated tools are used but its always human observation that is most important. The tools and instruments improve the accuracy when interpreted correctly. Now let's look at some complex sensors and sending devices.

6.1 Global Tools

A Meteostat weather satellite produces infrared images of earth's atmosphere. Several weather satellites orbit the earth, some provide both photographs and infrared images, and others carry radar and radiometers. An infrared radiometer measures temperatures at the top of clouds, while microwave radiometers see through the clouds.

Some modern technology acronyms are;

The Advanced Weather Interactive Processing System (AWIPS) is the integrating element of the National Weather Service Modernization Program. (http://199.26.34.19/AWIPS_home.html).

A total of 160 high speed weather computer and communications network sites are included in this system (http://199.26.34.19/site-TOC_index.html).

National Weather Service Mission: The National Weather Service (NWS) provides weather, hydrologic, and climate forecasts and warnings for the United States, its territories, adjacent waters and ocean areas, for the protection of life and property and the enhancement of the national economy. NWS data and products form a national information database and infrastructure which can be used by other governmental agencies, the private sector, the public, and the global community. (http://www.wrh.noaa.gov/pqr/).

The Automated Surface Observing Systems (ASOS): This program is a joint effort of the National Weather Service (NWS), the Federal Aviation Administration (FAA), and the Department of Defense (DOD). The ASOS systems serve as the nation's primary surface weather observing network. ASOS is designed to support weather forecast activities and aviation operations and, at the same time, support the needs of the meteorological, hydrological, and climatological research communities.

With the largest and most modern complement of weather sensors, ASOS has significantly expanded the information available to forecasters and the aviation community. The ASOS network has more than doubled the number of full-time surface weather observing locations. ASOS works non-stop, updating observations every minute of 24 hours a day, every day of the year. (http://www.nws.noaa.gov/asos/index. html). "The Automated Surface Observing System (ASOS) is an automated observing system being sponsored by the Federal Aviation Administration, National Weather Service (NWS) and the Department of Defense (DOD). ASOS provides weather observations which include: temperature, dew point, wind, altimeter setting, visibility, sky condition, and precipitation. 569 FAA-sponsored and 313 NWS-sponsored ASOSs are installed at airports throughout the country." (http://www.faa.gov/asos/asosinfo.htm).

Automated Weather Observing System (AWOS): This is a suite of sensors, which measures, collects, and broadcasts weather data to help meteorologists prepare and monitor weather forecasts. "There are over 600 AWOS sites located in the United States. The sensors measure weather parameters such as wind speed and direction, temperature and dew point, visibility, cloud heights and types, precipitation, and barometric pressure. The AWOS does not predict weather, but many send current information to weather offices where forecasts are produced using this information along with computer model outputs, satellite photos and radar images, to name a few." (http://www.faa.gov/asos/awosinfo.htm).

"There will soon be over 1500 FAA and NWS ASOSs, FAA AWOSs and non-Federal AWOSs. This total does not include the systems installed by the military or at private airports." (http://www.faa.gov/asos/hist-aos.htm).

NCAR = National Center for Atmospheric Research.

NCAR is a federally funded research and development center. "Together with our partners at universities and research centers, we are dedicated to exploring and understanding our atmosphere and its interactions with the Sun, the oceans, the biosphere, and human society." (http://www.ncar.ucar.edu/). "Although we don't issue official forecasts, our research is used by operational forecasters—those who issue regular outlooks for the National Weather Service, Federal Aviation Administration, military weather services, or private industry. We also help train the forecasters and design some of the computer models on which they rely." (http://www.ucar.edu/research/prediction/).

National Aeronautics and Space Administration (NASA)

On May 4, 2002, NASA launched Aqua, the sister to the EOS "flagship" — Terra — launched in December 1999. Aqua carries six state-of-the-art instruments to observe the Earth's oceans, atmosphere, land, ice, and snow covers, and vegetation, providing high measurement accuracy, spatial detail, and temporal frequency.

The purpose of NASA's Earth Observatory is to provide a freely-accessible publication on the Internet where the public can obtain new satellite imagery and scientific information about our home planet. The focus is on Earth's climate and environmental change. In particular, they hope the site is useful to public media and educators. Any and all materials published on the Earth Observatory are freely available for re-publication or re-use, except where copyright is indicated. (http://earthobservatory.nasa.gov/).

Aircraft

An airplane known as a Gulfstream IVSP flies at altitudes up to 45,000 feet (8.5 miles), drops windsondes (radiosondes) on parachutes to measure wind direction, and can study storms near the Troposphere.

Weather modification aircraft release seeding agents to enhance rain or prevent the formation of hail.

Hurricane hunter aircraft fly through hurricanes to collect data.

Weather balloons also carry radiosondes and usually rely on hydrogen gas to lift them as high as 100,000 feet. They carry instruments to measure temperature, atmospheric pressure, and humidity. By tracking the balloon with radar, information is gathered about wind speed and direction at different altitudes. Rain and snow scatter radio signals, so measuring reflected radio waves yields precipitation patterns. Thus we have radar imaging for rainfall.

Miscellaneous

- SKYWARN is a nationwide network of weather observers.

- Mobile Doppler radar trucks can be positioned in the path of storms.
- They are working on a S-POL radar capable of measuring raindrops.
- Storm spotting and tracking people are essential to the NWS.
- The National Hurricane Center (NHC) has a 110 year history.
- Public weather forecasting is 85% privately owned. Predominant in the private field are AccuWeather (with more than 10,000 clients) and,
- Weather Data (established in 1981). Weather Data is the inventor of SmartRAD which depicts storm warnings, wind speeds, lightning, temperatures, and precipitation type and intensity, all on the same computer screen. Some 300 companies exist to take up the slack if the NWS needs them.

6.1.1 NOAA

The National Oceanic and Atmospheric Administration (NOAA) currently operates 16 meteorological satellites in 3 separate constellations.

The National Polar-orbiting Operational Environmental Satellite System (NPOESS) and its managing Integrated Program Office (IPO) were established in 1994 to converge existing Air Force, NASA & NOAA polar-orbiting satellites into an integrated national program. Polar-orbiters in low-Earth orbit continue to be used to monitor global environmental conditions, collect, and disseminate data related to weather, atmosphere, oceans, land and near-space environment.

NOAA Satellites and Information Service (National Environmental Satellite, Data and Information Service) - NESDIS

This service operates the satellites and manages the processing and distribution of millions of bits of data and images these satellites produce daily. The prime customer for the satellite data is the NOAA National Weather Service, which uses satellite data to create forecasts for television, radio and weather advisory services. NOAA's operational environmental satellite system is composed of: geostationary operational environmental satellites (GOES) for short-range warning and "nowcasting," and polar-orbiting environmental satellites (POES) for longer term forecasting. Both kinds of satellites are necessary for providing a complete global weather monitoring system. The satellites carry search and rescue instruments, and have helped save the lives of about 10,000 people to date. The satellites are also used to support aviation safety (volcanic ash detection), and maritime/shipping safety (ice monitoring and prediction).

GOES = Geostationary Operational Environmental Satellite which measures clouds, water vapor, fire smoke, wind, temperature and ozone. An 'imager' provides the data on clouds, water vapor, fire, smoke, and wind. The 'sounder' collects temperature information and current ozone levels. GOES can zoom in on a specific severe weather event every five minutes. The GOES system maintains a continuous data stream from a two-GOES system in support of the National Weather Service requirements. These satellites send weather data and pictures that cover various sections of the United States. Current weather satellites can transmit visible or infrared photos, focus on a narrow or wide area, and maneuver in space to obtain maximum coverage. (http://www.goes.noaa.gov/).GOES 10 is the main one for our part of the world and GOES 11, which was launched May 3, 2000, is considered a spare. Their site shows 15 sectors of the world simultaneously and provides additional resources and information.

Polar Operational Environmental Satellite **(POES)**
The POES satellite system offers the advantage of daily global coverage, with morning and afternoon orbits that deliver global data, for improvement of weather forecasting. The information received includes cloud cover, storm location, temperature, and heat balance in the earth's atmosphere.

The future NPOES constellation will merge the two polar orbiting constellations into a single program.

6.2 Resources Conservation Service

Automation is the key to gathering large amounts of data and analyzing it to determine patterns over time with a high degree of reliability. As with Climate, trends in weather can be used to create a computer model with a high probability of accuracy. Once created, this software is used with current conditions to predict potential weather.

"The climate data currently used in Conservation Planning are generally observed by the National Weather Service (NWS) Cooperative Network. This nationwide network currently consists of nearly 8,000 active climatic stations. Observations at cooperative stations are performed by private citizens, institutions (such as utilities and television stations), or state and federal agencies. Of the nearly 17,000 NWS climate stations in the Natural Resources Conservation Service's Centralized Database System (CDBS), approximately 6,700 contain sufficient observation record length (greater than 20 years during the most recent normal period, or no more than 5 consecutive years of missing data) to provide representative averages and probability information. (http://www.wcc.nrcs.usda.gov/climate/).

The National Resources Conservation Service (NRCS) has the computer models for the Western United States on computers in Boise, Idaho and Ogden, Utah. At midnight the Boise computer contacts Galena Summit and some

600 sites in 10 western states submit data of current conditions. The remote sites are typically in small meadows in high mountainous country and many are accessible only by helicopter. Each site has a flat steel pillow filled with non-toxic antifreeze. Accumulating snow pushes the antifreeze up a plastic tube allowing measurement. Gauges and recorders measure temperature and wind using a small solar panel for power.

Radio wave transmissions bounce off a gaseous meteor trail to a single unmanned site which responds with a burst of weather data. Thank heaven for the constant shower of sand grain sized meteors that burn up in the atmosphere leaving a gaseous trail capable of reflecting a radio signal. Because billions of tons of cosmic dust fall to the earth each year, a signal is assured of randomly finding a short-lived trail for a quick burst of communications to bounce off. The data are temperature maximums, minimums, and averages, snowfall amounts, moisture content of the snow, and total accumulations of precipitation to date. That takes 50 milliseconds.

This data isn't just to forecast weather for your weekend outing. Runoff estimates from mountain watersheds can be forecast for farmers, hydroelectric power plants, outdoor enthusiasts, and others. The National Resources Conservation Service is gathering the data and making it available to many agencies and private enterprise.

6.3 Road Sensors for the Future

A high-tech road condition system does exist. Sensors can be embedded in pavement which, combined with weather instruments near roads, provide detail on current conditions. The sensors are metal plates mounted on thermistors. A

Thermistor is a device that changes its electrical resistance based on temperature changes. Metal pins can monitor salt and other chemical compounds on the surface. The electrical resistance varies for a limited range of chemicals. Cabled to remote processing equipment and the weather (wind, humidity, and temperature) station, better data is compiled to track changing road conditions. The technology exists but the lack of population density does not justify the expense.

6.4 Summation

The big guys (meteorologists) sure have some high tech tools to put together their weather prognostications. Regardless of all these devices, including the computer power available, it still takes a weather guy or gal to pull it all together for a useful report. You may notice the meteorologists are not trying to measure anything that even old Daniel Boone didn't try to measure; temperature, rainfall, air pressure, wind direction, wind speed, humidity and depth of snow.

What has happened is that, like the guys on the two mountain tops, we can see more of the planet the higher up we go, aside from our limited view out the window. So, we go from standing on mountain tops to see across the valley to much higher elevations, where satellites can see much more of the earth and its atmosphere. Also, the detail of the information such as size of rain droplets or air movement at high altitudes is infinitely more accurate. We can see the cloud patterns half a globe away. This helps on the West Coast as our weather comes from the ocean. Since weather travels from west to east on the East Coast they have more land based stations so it should be easier to forecast the weather.

The more recording stations go on line, the better the forecast. Sophisticated and specialized tools, with or without computer modeling, give people the ability to forecast better.

Larger amounts of data provide better trend data and global comparisons. Remember these professionals are worried about the world while you are forecasting right here at home on the Oregon coast.

Weather people can be somewhat general with their forecasts. The information provided will help you forecast what the weather will be where you are located. If you are going some place, believe the professional forecasters or, better yet, don't drive over that mountain pass at night if they forecast snow.

Start each day by making a guess about the weather. Pretty soon, with the ocean on one side and the coast range on the other, you will get a feel for the weather especially when you note the wind direction, usually from the North in summer and South in winter. The wind speed will give a clue as to how fast the weather will change. The faster the wind speed the quicker the change.

If all else fails get used to looking outside to see if your porch (or Porsche) is wet or dry. Is the sun shining? Are there clouds out there over the ocean? What is the wind speed and direction? You can make a pretty accurate weather guess this way.

7. STRANGE AND FUN WEATHER STUFF

7.1 Weather Glass

Early sailing ships needed a way to forecast weather changes 8 to 12 hours in advance. They used a simple glass filled with colored water. It has been called the thunder glass, storm glass, Amazing Weather Ball, or water barometer. This is a sealed glass container with a spout connected low in the bowl of the container. The water level in the spout rises with an oncoming storm and recedes when fair weather is on the way. The barometric pressure is pushing on the water so the small amount in the thin spout appears to react significantly and you can see changes in the water level representing higher or lower barometric pressure. Don't put it in direct sunlight unless you want to see the expansion of the water cause the water to bubble over the top. This would give you a false reading.

The weather glass dates back to 17th Century Netherlands. Historians believe the Pilgrims brought them to the colonies in 1620.

Reproductions are available in many gift shops around here but especially those shops dealing with marine items.

7.2 Vermont Weather Stick

The Vermont Weather Stick is a thin branch usually cut from a willow tree. Look at a willow tree and you will notice the branches are bending down. Turn that old willow stick around and let's have it forecast weather. The branch is mounted outside, under cover, and is used as a humidity indicator. If the humidity is high, it points down, and it curls up beautifully when the air is dry. Sure you could calibrate it with a more sophisticated instrument but that could take some of the fun out of it. The weather stick is about 14 inches long and, it is said, every farm house used to have one on the porch.

When the Vermont Weather Stick is bending down, this means higher humidity which is common on the Oregon coast. It is exciting when we have low enough humidity to cause the willow stick to point up. This does give a modicum of warning for weather changes. As good weather approaches the 'weather stick' will reach for the sky. When things aren't so good (assuming you don't want it to rain) it points to the ground.

7.3 Swiss Miss Style

One of our recent additions is the old time Swiss Miss and husband (we think it is her husband; they live in the same place). The fairy tale cottage they live in has a large front door. The instructions say place it in an 'airy' place where it will not be jostled. Turn the chimney until they are both in the doorway. In sunny weather the lady comes out and in rainy weather he comes out. It really works.

7.4 Weather Rock

Some of the simplest devices lend humor to forecasts. Consider the 'weather rock'. It does not give you much warning but it is very accurate. Once you have selected a site exposed to the weather but in view of the comfort of your home, you are ready to observe. Another advantage in the weather rock in contrast to other instruments is that it does not require batteries, adjustments, or plugging in.

It works like this. If it is wet it is raining; if it is dry it is NOT raining. If it is covered with snow, yep! When the snow starts to melt the temperature is going up. What could be simpler? The weather rock is a clever device and 100 percent accurate.

7.5 Donkey Tail

The donkey silhouette with a rope tail is a little more sophisticated weather indicator. The addition of the tail adds valuable information to your weather knowledge. The weather rock's concepts hold true here as well, but the tail gives us wind direction and an indicator of speed that no weather rock could hope to provide. Be sure to put the donkey in an exposed area not protected from the wind and where you can see it from inside your house.

It shows wind direction and speed. If the tail points straight down it is calm and there is no wind. Whichever way it's blowing indicates the direction from which the wind is going. If it flops around, it's breezy outside with variable wind direction. You must admit that is pretty clever, especially since, like the weather rock, you don't need electrical power.

7.6 Mahogany Case Barograph

This older instrument has a recording drum that rotates at a constant speed and can record from 7 to 31 days of information. The stylus is activated by an evacuated bellows like the one in an aneroid barometer.

They are as expensive as a new weather device, probably because of their antique value. Once a state of the art barometric recording device, it has been replaced by less expensive digital units. However, some of us would like to have one of these beautiful devices in our arsenal of weather gear.

7.7 Thermograph Recording Thermometer

The thermograph is housed in a classic case similar to the barograph. At first glance they look alike, but the ink stylus in this case is activated by bimetal strips which expand or contract differently with temperature variations. They can record temperature fluctuations over a 7 day period. Remember this instrument must be outside to record the proper temperature.

7.8 Galileo (Galilei) Thermometer

17th Century European glass blowers developed the Termometro Lento (slow thermometer) based on Galileo's principle that the density of liquid changes with temperature. The large tube is filled with a liquid that floats the glass globes.

Glass globes, suspended in liquid, descend as the temperature rises since the liquid becomes less dense and is not able to support their weight. This means each one slowly sinks to the bottom as the temperature rises. The last one in the cluster of globes is the temperature. The temperature increments available are based on the quantity of globes, generally about nine degrees each. Each globe has a temperature reading medallion usually hanging below it. Aside from being relatively accurate, the instrument can be purchased in beautiful glass form of various heights. This is a show piece for your house whether or not you are forecasting weather.

7.9 Sinuses

It isn't every day you meet a real life rain forecaster with a built-in system. A producer for one of Portland, Oregon's TV stations said she has an infallible rain forecaster built into her sinuses. A sinus headache is a rain forecast. The indication can be as much as eight hours in advance. Think about the process and it's not so strange. First, the humidity goes up just before it rains. Next, the barometric pressure goes down. So we have higher pressure inside the sinus passages that needs to equalize fast or you feel pain.

7.10 The Birds

Occasionally we get some other clues about extremes in temperature. It's hot if birds are in the shade and quiet, some of them holding their wings away from their bodies. They are cold if they are actively chasing around for food to fuel their bodies since the colder it is the more food they need. When they sit around all fluffed up and appear twice their size, it's to increase the warm air space next to their bodies.

People worry about one legged seagulls in their area. If you are patient you see it change legs. The gulls get cold feet and will hunker down and cover up both feet or stand on one leg to warm the other. We have observed gulls that have lost a leg but they seem to manage successfully.

Birds tend to perch more when a storm is approaching. The key is barometric pressure. Lower air pressure lowers density and thus less support for flight. So why strain, just perch as the storm approaches. When the gulls are all down in some park or open space, you can bet some heavy weather is coming in.

7.11 Animals

If you are lucky enough to have animals, the thickness of the winter coat may tell you a cold winter or a hot summer is on the way. No one has had a government-sponsored study to prove or disprove this, but it works. Ask any farmer about his horses or cows and their winter coats. When the cattle herd together and look for lower elevations the weather is probably changing for the worse.

Don't forget our house pets. They seem to react to barometric pressure; extreme highs and quick lows will alter their behavior. It has even been written that your cat may 'hear' the storm approaching.

Some people are lucky enough to live close to habitats that have frogs. Frogs increase their serenade several hours before a storm.

Perhaps you may also notice bees stay closer to the hive before a storm.

7.12 Your body

Those of you with built in barometers such as that trick knee or elbow get a warning of weather changes that the rest of us are willing to do without. The word is that a trick knee or elbow is strictly a reaction to barometric pressure changes.

The human hair reacts to changes in humidity. In 1771 Horace Benedict de Saussure discovered human hair is a good indicator of humidity. As the humidity rises hair gets longer (expands) and when the hair gets shorter (contracts) it indicates a decrease in humidity. Human hair absorbs water. When there is more water vapor in the air it expands the hair molecules. Weather can make hair appear thicker but can not alter thinning. This explains why naturally curly hair becomes frizzy and straight hair becomes limp with increases in humidity.

Did you know there is a hair hygrometer, which is an instrument to measure humidity? Credit goes to Leonardo de Vinci (1452-1519) for the first design. A form of instrument for plotting humidity changes is the hydrograph. As with the thermograph which uses two metals, a hydrograph uses a set of hairs under tension.

7.13 Plumbing

All plumbing fixtures that flush or drain water have a vent pipe open to the outside. If the pressure is low outside and your house is sealed fairly tight, the toilet bowl level will drop. The outside pressure on the pipe is lower than the inside pressure, so the inside pressure pushes the water down. We have never seen it move appreciably the other way and, we guess, this is because the three inch drain vent pipe can't get enough pressure in it to push up the gallon or so of water in the toilet bowl.

7.14 Odors

One reference suggests lower barometric pressure allows organic odors from swamps, marshes, or drainage ditches to rise up to the point we can smell rain. Normally the higher pressure keeps that old rotten egg methane closer to the ground; swamps smell like swamps. But a falling barometer means there is less air pushing down. As a result, smells of flowers, grasses, and such, are released. That may be the fresh aroma we experience prior to a summer storm.

7.15 Summation

Hopefully, you enjoyed our chapter on strange and fun weather stuff indicators as much as we did in pulling it all together. Keep in mind the indicators such as your sinuses or animal behavior have been used by humans for weather forecasting since we first arrived on this planet. The main difference between our current situation and the "olden" days is that if you did not make accurate weather guesses then, your life could be on the line. Then people relied on observable animal behavior for clues to weather changes.

Weather indicators like the weather glass or Swiss Miss Chalet are based on tried and true scientific principles. These devices use the air pressure or humidity to do their thing. The Vermont Weather Stick is a fun indicator because it does work. We didn't get into it, but check with your local gardening enthusiasts and they will show you diagrams of branch cell structure which explains how plants move. A case in point is the weather stick pointing up or down as the humidity changes. In some cases flowers follow the sun all day.

The main thing about these gizmos is they do not need batteries. You will be amazed how reliable the fun stuff is. However, these weather indicators don't appear to read as accurately as your digital weather station. This should not stop you from making daily observations with them. Their purpose is to make weather guessing fun and their ability to forecast trends is reasonably accurate. That has been our experience.

8. WEATHER GEAR

From your wanderings through Pogonips, you may now be ready to purchase some serious weather gear (instruments). It will not make your information any more accurate than does some of the fun stuff; however it looks more sophisticated and is a pleasant addition to your arsenal. Keep in mind what information you want for your weather guessing. You will want to know high and low temperature, wind speed and direction, barometric pressure (including direction of change), rainfall in inches or centimeters, humidity, and wind-chill.

There are two ways to go. Buy individual pieces so you can read and log the data, or buy new digital systems which will do it all, including downloading to your computer and creating beautiful charts and graphs. All this is done with a keystroke on the computer or a click of the mouse.

Some of us like to watch lights flash and needles move or mercury or the colored water going up and down in a glass tube. If you like this kind of fun you can get a combination of instruments; some digital and some analog.

Readout devices are inside but the weather is outside. Do you drill holes to route wires to probes and sensors or do you buy wireless systems? Wireless systems are newer and extremely reliable. Most of them rely on solar energy and transmit their findings to the base station. In either case the probes and sensors must be properly placed outside. Be sure to read the instructions that come with your weather station before installing the system. The temperature probe needs protection from direct sun and the wind. Solar shields are available to take care of the problem. Placing the probe up under cover on the North side of the structure is usually sufficient.

Many official weather stations use a shelter called a Stevenson screen or "weather shack" to protect the thermometer. It is a louvered box four feet above ground level.

8.1 Thermometer

Thermometers contain either alcohol or mercury which expands with heat or contracts when cooled.

Sure, a simple thermometer does the job for RIGHT NOW, but we are forecasting "experts". We need temperature over time. We need something to give maximum and minimum temperature readings. These can be mechanical or digital electronic devices. One device is a mercury column and some are U shaped. As the temperature goes up it pushes the mercury up the column and the mercury stays where it is until you push a button to release the mercury.

Consider a round dial thermometer. Yep, it too can give you maximum and minimum temperatures over time. The temperature indicating needle bumps into pointers on its way up or down. The farther these needles are pushed the higher or lower the temperature has been since the last mechanical reset.

If you go digital, these instruments record the highs and lows accurately, but you will need to reset them too.

How do they work? Who cares? These are great because the readout is inside. Read the instructions and have fun. We have one that has a pleasant female voice hourly announcing the temperature inside and outside. She will announce a warning if the temperature drops to freezing. Again a simple push of a button usually announces the maximum and minimum since last reset.

8.2 Wind

Direction and speed is what you want to know. Some instruments will indicate wind direction with red lights; others will indicate direction with a compass rose. Wind speed may be displayed by a gold needle and a black needle. The gold needle is highest wind speed and the black needle is the current wind speed. Another type of station indicates wind speed (maximum and current) in miles per hour, knots, or kilometers on a readout device. A simple system allows you to read the highs and lows or another system passes the information to the 'computer' for future use. We feel the old fashioned way of charting this data on our own charts is more fun.

You will want to know what the wind speed is and there are many ways to measure it. All we need is something that reacts to wind. SURE! Use a windsock, a string, a weather vane, moving branches, or anything that will give some indication of speed.

The instrument may capture the wind and push a column of mercury up a tube, straighten out a windsock or just bend the branches, but what we are really looking for is something that can be calibrated and gives reliable indications.

Thus enter the three cups rotating on a shaft and sending electrical impulses proportional to the speed of rotation. These electrical signals are converted to analog or digital indicators for our use. This anemometer (wind speed indicator) needs to be away from other buildings or parts of the house that may block or redirect the wind. The anemometer is omni directional; as long as a breeze can get to it, it will rotate.

Frequently a wind direction indicator is part of an anemometer setup. The wind direction element needs a clear shot at the wind to be able to swivel 360 degrees. As a hobbyist you can get all kinds of weather vanes to indicate direction. You can spend as little or as much as you want on weather vanes. Some are not only functional wind direction indicators, but they also are works of art to enhance your house or yard. The simplest ones are as low as less than $10, with expensive ones as high as $300.

Airports usually have a windsock but they have also used another device called a wind T. It is a T shaped device that looks like an airplane with no tail. It is pretty good size too. Some are free to pivot in a concrete circle that has a compass rose on it. Because the wind tee is similar to an airplane, it will rotate like a weathervane into the wind. When the wind isn't blowing enough to move it someone manually points it in the direction. This shows the pilot how to approach the runway. He flies in the direction of the long end of the T just as if it is the airplane in front of him. This is for a landing into the wind. The wind T is not as common as the wind sock. However, it is still effective in helping determine the wind direction and favored runway. Depending on the wind speed a pilot may be directed to make a down-wind (with the wind) approach.

8.3 Atmospheric pressure: the Barometer

A French scientist named Blasé Pascal (1623-1662) was the first to note changes in atmospheric pressure relate to changes in the weather.

High and low pressure is always relative. There is not a number that divides low and high atmospheric pressure. In general, low atmospheric pressure at the surface means clouds and precipitation, while high atmospheric pressure usually will indicate clear skies.

When talking about the direction of change of the barometer (up or down) we are attempting to predict what will happen next. Remember, the direction of change of pressure is important. Rising pressure could be unchanging (settled) weather while a drop in atmospheric pressure signals unsettled weather. For certain, the weather will change very soon if is not already doing it.

Now let's discuss measuring atmospheric pressure. The mercury barometer was invented by a pupil of Galileo in 1643; Evangelista Torricelli (1608-1647). It was Galileo who suggested Torricelli use mercury in his vacuum experiments. Torricelli filled a foot long glass tube with mercury and inverted the tube into a dish. Some of the mercury did not escape from the tube and Torricelli observed a vacuum was created.

Today's instruments use the same basic mechanism. A simple tube, closed on one end, is almost filled with mercury and placed open end down in a pool of mercury. A ruler is placed next to the tube. If pressure on the pool of mercury rises, it causes mercury in the tube to rise and visa versa.

There were other variations of the original tube barometer which have not stood the test of time. Robert Boyle (1627-1691), an Anglo-Irish scientist produced two models--a water barometer and a more portable siphon barometer using a U shaped tube filled partially with mercury.

Another application is the wheel barometer invented by Robert Hooke (1635-1703). Hooke was a colleague of Boyle. The device is J shaped with a float in the bottom of the J resting in a pool of mercury. The upper leg of the J was enclosed. The upper leg also contains mercury which, when it moved, created a vacuum. As the mercury moved so did the float attached to a cord around a wheel. A counterweight on the other end of the cord provided the necessary delicate balance. As the wheel was turned by the cord, an attached pointer showed barometric changes.

We measure atmospheric pressure in inches of mercury. When you hear pressure in millibars it is the metric equivalent of inches. Kilopascals (millibars divided by 10) are another measure. You need about a 32 inch tube of mercury to capture the full range of pressure change. This is rather inconvenient because of its size. Thus the aneroid (without liquid) barometer came into existence. The person credited with the invention of the aneroid barometer in about 1843 is Lucien Vidie, a French scientist. Now we can measure atmospheric pressure in pounds per square inch.

The aneroid barometer is the most widely used instrument for atmospheric pressure measurement. It is a flexible metal bellows (usually steel or beryllium), tightly sealed after having some air removed. Like a tiny accordion, higher outside pressure squeezes the bellows to which a pointer is attached.

The old round dial aneroid barometer was once a popular special occasion gift. Many of us had one, looked at it admiringly, and then ignored it since not much happened; no bells or whistles to hold your attention. If you get in the habit of recording readings daily or during a storm you will see it really has lots of action.

There is a knob on the front of the aneroid barometer which moves the usually gold colored needle. Turn it until it is directly over the indicator needle. Once you have done this, tap the glass of your barometer and the indicator needle will stay either where it is, or move. Note the numbers usually go from 29.00 to 32.00. Generally 30.00 inches is in the middle of the indicating range. At or above this range usually indicates calmer weather. Below this range means a low pressure and a change in the weather commonly called deterioration. By logging daily readings you will note the same general idea holds true above and below 30.00 inches.

On the back side of the aneroid barometer is a set screw to compensate for altitude, since higher altitudes have lower air pressure. Find one of those 'official' stations nearby; Coast Guard, NOAA at 162.550 MHz, Weather Bureau or just follow local weather reports, noting time of their report and comparing it to your instruments reading. This will allow you to average out your barometer setting. Calibration is important. Sun, heat, or drafts affect the pressure reading. These instruments also tend to drift over time, so frequent calibration may be necessary.

At sea level the force pushing against each square inch of any exposed surface is 14.7 pounds. Remember this pressure is equal in all directions; up, down and sideways. Because atmospheric pressure decreases with increases of altitude, aircraft use a special aneroid barometer called an altimeter. A change in atmospheric pressure of one-tenth of an inch of mercury will make an altimeter reading change 100 feet. Aircraft must contact the airport for a local pressure correction to adjust the altimeter for current barometric conditions on the surface.

8.4 Rain Gauge

The rain gauge is probably the oldest meteorological instrument. Rain gauges are of several types. Most are designed to give us inches of rainfall. One inch of rainfall is equivalent to a layer of rain one inch deep on level ground. Tipping bucket gauges have a divided bucket that fills one side, tips its water out, and then allows the other side to fill. This gauge can be so precisely balanced that 1/100 inch of rain tips the bucket each time. As the bucket tips it sends an electrical signal to a recording drum or sophisticated computer chip inside the weather station.

Another device continuously weighs the water, recording periodic measurements (weights) as inches of rainfall.

There are digital ones that read in tenths or hundredths of an inch.

The easiest and cheapest one is a plastic cone that is stuck in the yard or mounted on the deck. Some are designed to be read from a distance so you can sit inside, warm, and dry. Some rain gauges are self emptying; others you need to empty and record the data. Remember you need to log your readings over time to compare with 'official' reports to establish confidence in your own system.

We consider precipitation information extremely important. The amount of annual rainfall and its seasonal distribution determine the kinds of plants that will thrive on our coast.

8.5 Humidity (psychograph or hydrograph)

The humidity gives clues to weather changes. Humidity information is valuable because humidity affects your well being. For weather guessing you need the trend of the humidity, not the exact percentage.

Talking about relatives? Yes, relative humidity is the factor and the sling psychrometer was the first instrument used to measure it. It is actually two thermometers, one of which is a regular mercury thermometer like the one your household has. The second thermometer is a wet bulb thermometer because it has a muslin wick over the bulb that is dipped into water before each reading; other than the wick the two thermometers are identical.

To get more exact readings outside, spin the thermometers in the air. Whirling the two thermometers causes water to evaporate from the wick which lowers that thermometer's temperature. You need a table (chart) to turn your readings into humidity percentages.

Consider the current weather. If it is dry obviously more evaporation will take place around the wet bulb thermometer's wick and thus more cooling. The difference in temperatures of the two thermometers is used to measure the relative humidity from a table constructed for these instruments. The operator reads both temperatures and looks at the chart intersection to obtain the relative humidity. The amount of water vapor (humidity) the air can contain is directly related to the temperature of the air.

Instruments that read humidity on a dial (analog or digital) are based on an invention that goes way back. An advanced version that the big guys use is called a hydrograph. It uses two or more sets of hairs under tension that move a stylus in a drum. Some instruments use a sheaf of blonde human hairs treated to remove oil. As the relative humidity increases the hairs increase in length. The same principle as used with the thermograph or barograph.

Digital instruments involve a sensor that changes electrical properties based on the humidity. This change is converted electronically into a digital display.

8.6 Dew Point

We have already defined the dew point as the temperature at which the air is saturated. The air contains as much water as it can hold without it precipitating out. Humidity is said to be 100 percent so the dew point temperature and the current outside temperature are essentially the same. If the temperature is at its dew point and then drops as much as one degree precipitation will occur.

What instrument provides you the dew point? The outside temperature thermometer and a humidity gauge give you the related factors. As defined in section 4 of this book, use that data with a formula to compute dew point. A better way is purchase a digital weather station which will provide a direct readout of dew point.

8.7 Summation

Considering this chapter on Weather Gear and the one just before it, we have gone from the ridiculous to the sublime. All of it is fun for the amateur. All of it is with some basis in meteorological science.

The advantage of a complete digital weather system, its software, and a printer is the ease of keeping records. This is in contrast with thumbing through pages of handwritten information.

Individual weather gizmos (toys) should be included regardless of your state-of-the-art equipment. Despite battery backups, etc. a redundant system of manually read instruments keeps you from having breaks in your data especially when the fancy stuff goes in for repairs. Individual gizmos discussed previously are fairly error free and reliable. Digital systems have been known to crash. It also helps to keep your eye out for a subtle decline in accuracy for one reason or another. If you are like us, you are trained to not always believe your instruments so create another way to cross check the data.

Once your friends find out you have Weather Gear, be prepared to answer all kinds of questions. Our advice is not to get too good a reputation or your friends will be asking you if they can have that outdoor barbecue. Heaven forbid if it rains on your prediction of clear and sunny. However, you can always point to the fact the big guys also goof.

9. WEATHER PHENOMENA

9.1 Destructive Winds

When the wind blows your fence down or turns your umbrella inside out or you are stopped from driving across the Newport bridge, it really doesn't matter what you call it, wind is wind. Wind is a bunch of air that is in a hurry to get from one place to another.

Higher air pressure pushes air toward lower pressure. For example, if one area has 15 pounds per square inch of air pressure and another area 500 miles from us has only 14.5 pounds per square inch of air pressure, the air movement would be accelerated to 80 miles per hour in three hours (given no other forces are applied). Double the distance between these pressure gradients and the wind speed is one half (40 mph).

Hurricanes are special and they are probably the earth's most awesome storms. Typhoons, hurricanes, and tornadoes are some terms that show up on weather charts. "Our civilization is affected by these howlers ranging from hurricanes, typhoons, cyclones, and storms which belong to the family of tropical cyclones." (http://www.typhoon2000. ph/TCGuide.htm).

"Indeed, the term cyclonic not only means that the fluid (air or water) rotates in the same direction as the underlying Earth, but also that the rotation of the fluid is due to the rotation of the Earth. Thus, the air flowing around a hurricane spins counter-clockwise in the northern hemisphere and clockwise in the southern hemisphere (as does the Earth, itself). In both hemispheres, this rotation is deemed cyclonic. If the Earth did not rotate, the air would flow directly in towards the low pressure center, but on a spinning Earth, the Coriolis force (discussed below) causes that air to be deviated with the result that it travels around the low pressure center." (http://www.ems.psu.edu/~fraser/ Bad/BadCoriolis.html)

The terms **"hurricane"** and **"typhoon"** are regionally specific names for a strong **"tropical cyclone"**. A tropical cyclone is the generic term for a non-frontal synoptic scale low-pressure system over tropical or sub-tropical waters with organized convection (i.e. thunderstorm activity) and definite cyclonic surface wind circulation <u>(Holland 1993)</u>. (http://www.aoml.noaa.gov/hrd/tcfaq/A1.html).

It depends on where you are in the world. The Caribbean residents call the storms hurricanes, while in India and China the same type of storms are generally called typhoons. They swirl counter clockwise north of the Equator making big pinwheels on the weather charts shown as clouds. Once the tropical cyclone reaches winds of at least 17 m/s (34 kt, 39 mph) they are typically called a <u>"tropical storm"</u> and assigned a name. If winds reach 33 m/s (64 kt, 74 mph)), then they are called:

"hurricane" (the North Atlantic Ocean, the Northeast Pacific Ocean east of the dateline, or the South Pacific Ocean east of 160E)

"typhoon" (the Northwest Pacific Ocean west of the dateline)

"severe tropical cyclone" (the Southwest Pacific Ocean west of 160E or Southeast Indian Ocean east of 90E)

"severe cyclonic storm" (the North Indian Ocean)

"tropical cyclone" (the Southwest Indian Ocean)

<u>*(Neumann 1993)*</u>. http://www.aoml.noaa.gov/hrd/tcfaq/A1.html).

Hurricanes and typhoons pride themselves in indiscriminately taking out large territories of helpless people instead of a narrow path like tornadoes.

A tornado is "a violently rotating column of air, hanging from a cumuliform cloud or underneath a cumuliform cloud, and often (but not always) visible as a funnel cloud". Literally, in order for a vortex to be classified as a tornado, it must be in contact with the ground and the cloud base." (http://www.spc.noaa.gov/faq/tornado/#The%20Basics).

Tornadoes are concentrated, so a quarter-mile wide tornado is huge, while a 100 mile wide hurricane is possible. The tornado decides to be weird and spin so tight it looks like a funnel dancing across the landscape, playing a dangerous roulette game with a few poor souls in its path. Tornadoes can last from several seconds to more than an hour. The word gale fits in here someplace. No matter what the wind is called, it does an impressive job of tearing up things. Winds in the strongest tornadoes have topped 300 miles per hour while hurricanes rarely reach 150 miles per hour.

Tornadoes and hurricanes have another trick up their sleeve. Since it is circular, the wind can hit from one side and acts as though the storm has ended. In fact, the eye or center of the storm may be calm and dry. Just as you relax it hits you again from the other direction as it passes through. Tornadoes seldom last over an hour while hurricanes may easily last for days. The difference is the size in square feet or miles of the 'funnel'.

A large hurricane stirs up more than a million cubic miles of the atmosphere every second. Its winds can produce 50 foot waves or even higher in the open ocean. When this water hits the shore it may peak at over 20 feet and flood 100 miles of coast. A hurricane dumps six to twelve inches of rain when it comes flooding ashore.

We aren't sure which is worse; earthquake or wind. We try not to complain about winds too much, since a blizzard and subfreezing cold with yards of snow on top of everything is another of nature's tricks, which we seldom see here on the coast. The only justice is no matter where you go some weather extreme can make life miserable.

9.2 Sun Spots

Sun spots are large black spots on the sun. Sun spots have been known to exist from reports dating back to the first century. Since the 1870s people have been trying to establish links with them and weather changes. The issue is what do they do to the weather? It's known that sun spots increase and decrease in size on an eleven year cycle. Large spots appear when the sun's energy output increases. Smaller spots mean less energy radiated to the earth.

There was a cold spell known as the little ice age in Northern Europe during a low sun spot period. Looking for a link between sun spots and the weather makes sense since the sun drives the weather. So far, existing correlating methods have not been able to establish a cause and effect relationship. The best explanation for this lack of measurable cause and effect is the weather is too complicated. You can't quantify all the factors that come into play.

Sun spots may cause an overall effect in the climate (except for the little ice age confined to Northern Europe) but this is not a reliable prediction for weather changes. Remember climate is the weather over long periods and weather is the day to day events.

9.3 Rainbows, Halos, and Corona

There are a series of light displays that occur in the sky under special conditions. Some are more prominent in specific parts of the world. These displays are sun dogs, haloes, coronas, and rainbows. Rainbows are one display we frequently enjoy on the Oregon coast.

Most people have seen a rainbow, but not too many of you have seen the pot of gold that is always at the end of the rainbow. On the beach, often we get lucky and see a complete arched rainbow since the view is unrestricted over the ocean. All rainbows make an arc of 42° that require a wide angle camera lens to capture.

We will not bore you with a detailed technical analysis explanation, but in easy terms water droplets bend the sun light that is peaking through the clouds behind you. The bending is called refraction and it separates white light into the colors we see.

A sun dog is a bright spot visible on both sides of the sun or as a halo radiating around the sun. It is said that if the sun dog is on the right side and close to the sun, expect a southeast wind and nasty day. Likewise, if the sun dogs are on the left side of the sun and very close, there will be a northeast wind and again, nasty weather. Here on our beach you may not see sun dogs since it requires lower air temperatures than we usually have. It's worth noting however, because in this case, light is reflected from ice crystals rather than being bent or refracted.

If conditions are right you will see haloes. Halos are a white or faintly- colored ring either around the sun or the moon. They are hazy circles of colored light reflecting off ice crystals that are free-falling or suspended in cirrus clouds. The light may appear as a faintly colored ring with red near the center and blue on the outside. We have seen them around the moon, but can't remember seeing them around the sun. Some people suggest this means rain but scientific evidence doesn't support this. It does mean there is enough moisture to form ice crystals.

Coronas are caused by slight bending of light passing through clouds. The name means crown. The corona's colors are reversed from halos because of the difference between refraction and diffraction; bending versus reflecting. Each color has a different frequency so it bends at a different angle and we see some of the individual colors.

Suffice to say haloes require ice crystals and coronas require water droplets. Rainbows remind us that different colors are bent at different angles. The primary colors are separated along the spectrum. Don't let scientific explanations stop you from enjoying the light shows on the beach.

Just in case you may be interested, there is a light phenomenon called the pilots halo. In the old days, when you flew in piper cubs, once you got high enough to be above the clouds you might see a halo with rainbow colors and a cross in the middle following you. Bush pilots say it brings good luck. It did for one of us during a flight over St. Lawrence Island in Alaska; but that is another story.

The reason for the pilots halo seems to be a combination of things. Light bending around the airplane diffracts and a combination of reflecting and refracting causes the different colors. Enough said on that.

9.4 Green Flash

No, your neighbors are not imbibing too much if they call you all flustered about seeing a green flash. The green flash is a rarely seen natural phenomenon that happens at sunrise and sunset. At sunrise the disk of the sun is preceded by the brilliant green spot. It immediately vanishes into sunrise and it is rather difficult to see from our beach with hills to the East.

The best chance to see the green flash is when the sun sets; with the right atmospheric conditions. As the sun plops into the ocean a green flash appears, but you had better not blink. Just as the sun sinks a flash that is distinctively green will happen so quickly you better not blink. West of Lincoln City the sky needs to be cloudless and the horizon a fierce red just before it sets.

Get used to watching on those perfect sunset evenings, but be careful not to burn your retina by staring wide eyed as the sun goes down. At sunset the flash occurs just after or just as the sun sinks below the horizon.

What is really going on? This is really a lighthearted approach since the scientific explanation takes all the fun out of it. As the sun sinks the light rays bend (atmospheric refraction of light) and separate out, similar to how a prism works. As it sets the colors get blocked out one-by-one with green being the last color to pass through more of the atmosphere than when the old sun is high above the horizon. Refraction is stronger for greens and blues so the colors appear 'lifted' a bit, so briefly, the flares are left to startle the viewer. The more 'junk' in the atmosphere the more likely you are to see the 'green flash.'

In case your friends don't see one, you can use our book to prove you are not nuts. It really does happen.

9.5 Coriolis Effect

It was Gustave-Gaspard de Coriolis (1792-1843) who discovered the earth spinning deflects the streams of air. I'll bet nothing will be named after us Beach Weather Watchers. Maybe that's because Kohl and Jones do not sound as scientific as Gustave-Gaspard de Coriolis. It was 1835 when Gustave-Gaspard de Coriolis first explained the Coriolis Effect. It is the winds you see on television weather shows that curve across the northwest. These winds are also called trade winds.

Warm moist air at the equator rises until it stops then cools and then flows back down at about 30 degrees North and South latitude, back to the equator. Lincoln City is about at the 45th parallel, about halfway to the pole.

At approximately 60 degrees north and south this warm moist air meets the cold polar air and we have polar fronts. The temperature differences of the two fronts create Ferrel cells. William Ferrel first identified them in 1856. George Hadley (1686-1768) first described the flows these fronts create in 1753. They are now known as Hadley cells causing high pressure in dry desert regions.

Enough about the related air flows, now for coriolis. The amount of deflection the air makes is directly related to both the speed at which the air is moving and its latitude. Therefore, slowly blowing winds will be deflected only a small amount, while stronger winds will be deflected more. Likewise, winds blowing closer to the poles will be deflected more than winds at the same speed closer to the equator. The Coriolis force is zero right at the equator.(http://ww2010. atmos.uiuc.edu/(Gh)/guides/mtr/fw/crls.rxml). The Coriolis force turns the wind. Winds and thus weather systems turn to the right in the Northern Hemisphere, and to the left in the Southern Hemisphere.

Recall we said Coriolis Effect not force. What is the difference? The earth's rotation creates a force and mathematically it has an effect. **Coriolis effect** (for Gustave-Gaspard de Coriolis, a French mathematician), tendency for any moving body on or above the earth's surface, e.g., an ocean current or an artillery round, to drift sideways from its course because of the earth's rotation. (http://www. infoplease.com/ce6/weather/A0813558.html).

If you looked at winds from outer space they would appear to move in a straight line, but here on earth we experience the wind curving around our spinning globe. As we have already stated, obviously there is no coriolis effect at the Equator, and it is most pronounced at the poles.

9.6 SOI - Southern Oscillation Index

The Southern Oscillation Index (SOI) is a 'normalized' inter-annual see-saw in tropical sea level pressure between eastern and western hemispheres. "The Southern Oscillation Index (SOI) is calculated from the monthly or seasonal fluctuations in the air pressure difference between Tahiti and Darwin." (http://www.bom.gov.au/climate/glossary/ soi.shtml). The SOI provides a measure of trade wind strength. Temperature oscillations in the southern waters allow identifying El Niño conditions (warm waters) or La Niña (colder waters).

The SOI will be high for a La Niña which indicates colder than average winter temperatures. This condition appears to hold true for both monthly and seasonal averages and many short-term (day) averages. Growing seasons are longer for El Niño times (summer SOIs) predominantly because of milder late winter temperatures and early spring-like temperatures.

A study completed in 1991 determined the Southern Oscillation Index can be used for climate predictions, especially in winter. A four month lag time occurs in winter, so a summer average SOI correlates well with conditions in the Northwest during the following winter. SOI values less than zero represent El Niño (warm) conditions with 'near zero' values considered normal. The El Niño-Southern Oscillation (ENSO) exerts a strong influence on global weather and climate patterns. A negative SOI means a high El Niño; conditions of a mild winter. For the most part, El Niño conditions correlate with below average precipitation the following winter.

Oregon Climate Service (OCS) studies various aspects of the SOI-climate relationship. In southern Oregon the correlation is rather low but north of Roseburg the correlation is fairly strong. Oregon and the Pacific Northwest are strongly influenced by El Niño-Southern Oscillation. A consistent correlation exists throughout Oregon between SOI and total snowfall. If it's a strong El Niño, total snowfall in valleys is relatively low both east and west of the cascades. Moderate (near zero) SOIs also may have low snowfall but the positive SOIs are years with greatest snowfall occurring.

Hold tight. Don't give up on us beach weather watchers yet. We just stumbled on a paper by the Oregon state climatologist that have some clues for forecasting the winters here on the coast based on ocean temperature. The ENSO and SOI contain the secret. We both like the way the big guys come up with acronyms that are so classy. They compute values for these temperature oscillations in the southern waters which allow them to identify El Niño or La Niña. They also establish a four month lag before the weather reacts to the southern oscillation.

The question remains, did the ocean change the atmosphere or did the atmosphere affect the ocean. Anyway, let's get to what happens to us along our beaches. The 1997-1998 El Niño winter years on the Oregon beach were warm but rain was close to normal. In a La Niña year the coastal weather is wetter and colder than normal. The reason the 1995 ocean temperatures were close to normal was the La Niña effect. In 1998 the ocean temperatures were again close to normal (50-56 degrees). Say we are going into a La Niña year. Will it be wetter and colder?? The big guys will be the first to admit they haven't tracked enough cycles to bet the farm on the concept.

9.7 El Niño

El Niño (increase in sea temperature) is Spanish for Christ Child. The term was originally used by fishermen off Peru to refer to a warm ocean current that typically occurred around Christmas. El Niño is difficult to explain, since it's discussed interchangeably with cause and effect. Sustained negative values of the SOI often indicate El Niño episodes." (http://www.bom.gov.au/climate/glossary/soi.shtml).

El Niño seems to be responsible for all kinds of other things; fewer hurricanes in the Eastern Caribbean, more Hurricanes in the Southwest Pacific, more droughts in Indonesia, increased rain in the Southern United States, droughts in the Northwest, changes in the fish populations, and drastic die-off of marine birds and mammals. Biologists say the warm, sterile water result in ocean conditions very harmful to salmon.

Which comes first, the chicken or the egg? Did the trade winds stop blowing, so ocean currents which allowed the sun's radiation to warm the water in South Center Pacific? Did the water warm up stopping the trade winds which stopped the upwelling? This is an oceanography process in which cold nutrient-rich water rises to the surface from the ocean depths. Did the upwelling cease which warmed the surface water which reduced the trade winds which allowed the solar radiation to increase the temperatures of the oceans even more?

The mechanism for explaining how the waters warm up is not clear. It is less apparent what impact these temperature increases have on the dynamics of the system. Some suggest thermal action in the ocean starts the event. Hot volcanic vents in the ocean and magma intrusions are possibilities. At this time there is no evidence of increased plate action making more rift magma available. Solar radiation is another possibility with a change in cloud cover and cooling winds. It's most likely a combination of events that tips the balance and causes the observable phenomena.

To further explain El Niño, the ocean temperature rises, east and west temperature differences in the South Pacific Ocean are reduced, and the trades weaken. The warmer water, which has been pushed westerly, moves easterly and continues on to the coast of Peru and Ecuador. If the mass of warmer water continues to flow easterly it will move up the coast as far north as Washington State. There are other things happening such as the jet stream dropping southward and more storms moving into California and across through Texas.

Once this takes place the effects are wide spread. Some people are happily enjoying unseasonably good weather. Other people are getting needed rain and snow instead of drought conditions.

As was said before, the precise cause and effect will not be known until more data is collected and analyzed. What we do know is how necessary it is for man to review his effects on the global climate. At one time there was no way any group could affect global patterns. We know now world-wide logging, industrial pollution, city smog, and agricultural practices do make changes and add to the imbalance that fuels the whimsy of Mother Nature. Time will hopefully provide more definitive proof.

9.8 La Niña

Most people are familiar with El Niño, now that it has been identified by the world climatologists as a definite pattern of warmer waters. The opposite of El Niño is La Niña or a cold pool of water entering the same areas that were occupied by El Niño waters. Positive values of the SOI are associated with stronger Pacific <u>trade winds</u> and warmer sea temperatures to the north of Australia, popularly known as a <u>La Niña</u> episode"

(http://www.bom.gov.au/climate/glossary/soi.shtml).

La Niña, "the little girl", means cold ocean temperatures in the Equatorial Pacific. Another phrase for this is El Viejo (anti-El Niño). El Viejo is defined as a cold event or cold episode.

The phenomenon is currently being added to the weather patterns. Its effect, in simple terms, causes U.S. weather patterns to change in ways similar to El Niño. However, various sections of the country most likely will experience less wind, rain, snow, etc. If that sounds like double talk, see if this helps. The beach gets rain, wind, snow etc. during either event. However, during El Niño the northwest usually is dryer and warmer during the winter. This can vary, but the pattern is there. During La Niña the weather is rainier and colder with a good possibility the snow and the dry spell in summer and fall are not as severe.

The dynamics of La Niña have not been documented or tracked as precisely as El Niño, but with the installation of more weather buoys, the data is being compiled and past records back to 1950 are being analyzed. In July of 1998 the first La Niña conference took place in Boulder, Colorado at the National Center for Atmospheric Research.

Strong La Niñas do not automatically follow an El Niño event. The last strong La Niña was after the 1986-87 El Niño. Weather data shows increased rain and snow on the beach at that time. While La Niña patterns have not been analyzed as carefully, the pattern seems to be holding true for fall and winter conditions.

9.9 Lightning

Here on the beach and generally in Oregon, we are not zapped by lightning. Compared with the Midwest and other regions of the country, we see fewer electrical storms. The reason is the temperature differences between the land and the ocean are not as dramatic, so less turbulence is generated in air masses.

However, if you have been on a golf course or had your favorite trees split apart or lost something or witnessed a loss by a strike, you will doubtless disagree. Conditions on the beach are different. Over the ocean there are no high points such as trees or buildings-only boats. What has that to do with it? The way lightning works is as follows. It is sure more complicated than we first thought. In fact texts say there are two theories. The operative word is theory, which means the 'big guys' are not sure.

Benjamin Franklin established that lightning was an electric spark. Franklin was able to infer that while the clouds were overhead, the lower part of the thunderstorm was generally negatively charged. One theory is based on precipitation and suggests that smaller size rain drops, graupel and hail, may turn into positive charges, while the heavier ones move to the bottom of the cloud and are mostly negative. Second, is the convection theory that suggests upward convection currents of air carry positive charges from the ground to the cloud. Down drafts carry negative charges downward in the clouds.

The pathway for a strike is from the ground up. Set in motion by negative charges that form at the bottom of the clouds and a positive charge near the ground. Build up of a difference in strength strong enough for a spark to jump across the gap between the cloud and the earth. Negative charges are zigzagging downward until they near the ground. This draws a positive charge upward; the 'leader.' This leader stroke we do not see but it sets up a low resistance path for the return stroke of an intense wave of positive charge at a speed of about 60,000 miles per second (one-third the speed of light). This is lightning.

Now let us say that again only slightly differently. Negative charges (electrons) zigzag downward in a forked pattern. This is called a stepped leader.

As the stepped leader nears the ground it draws the positive streamer (charge) upward. As the leader and the streamer come together, powerful electrical current flows. Contact between the negative and positive charges is made, and an intense positive charge travels upward. This travels about one-third the speed of light and is the lightning we see. If you see a flicker it is because the process is repeating along the same path in less than one-half second intervals.

The awesome power and speed of a 'strike' is apparent to anyone that has seen lightning flash. "A lightning flash is composed of a series of strokes with an average of about four. The length and duration of each lightning stroke vary, but typically average about 30 microseconds. (The average peak power per stroke is about 10^{12} watts.)" (http://thunder.msfc.nasa.gov/primer/primer2.html).

The resulting thunder clap is easier to understand. The bolt of lightning parts the air and sound is generated along the length of the lightning channel as the atmosphere is heated by the electrical discharge to the order of 20,000 degrees C (3 times the temperature of the surface of the sun). (http://thunder.msfc.nasa.gov/primer/primer2.html). When the air comes back to fill the hole (and cools rapidly) you get a rapid air expansion/contraction which results in the sound we call thunder.

Light travels at 186,000 miles in a second, almost a million times the speed of sound. Sound travels at the relatively snail pace of one-fifth of a mile in a second, so the light versus the sound is a measure of how far away the 'strike' is. For fun, you can tell how far the strike is from you. When you see the flash the sound comes later. Now, count the seconds between the flash and the thunder boom. Count slowly!

Example: one-one thousand, two-one thousand, three-one thousand to approximate one second for each count. Count from the flash of light until you hear the boom. Five seconds is roughly equivalent to one mile in distance.

Cloud-to-ground lightning is the most damaging and dangerous form of lightning. Intra-cloud lightning is the most common type of discharge. Inter-cloud lightning, as the name implies, occurs between charge centers in two different clouds with the discharge bridging a gap of clear air between them. Details of why a discharge stays within a cloud or comes to ground are not understood.

Let us talk about some general safety rules on the beach in case of lightning. When caught out in the open, stay away from tall things. Avoid poles and trees. Generally anything projecting above the landscape is a source for the electrical current flow. Do not stand on a hill top. Your car is a relatively safe place provided you consider keeping it amongst other vehicles or buildings.

Get off and away from open water. If you are in a group in the open, stay several yards apart.

Feel like your hair is standing on end? DROP TO YOUR KNEES; do <u>not</u> lie flat on the ground. That is the theory but we have not personally tested that hypothesis.

There are numerous names and descriptions of various types and forms of lightning. Some identify subcategories, and others may arise from optical illusions, appearances, or myths. Some popular terms include: ball lightning, heat lightning, bead lightning, sheet lightning, silent lightning, black lightning, ribbon lightning, colored lightning, tubular lightning, meandering lightning, cloud-to-air lightning, stratospheric lightning, red sprites, blue jets, and elves.

"Lightning, the thunderbolt from mythology, has long been feared as an atmospheric flash of supernatural origins: the great weapon of the gods. The Greeks both marveled and feared lightning as it was hurled by Zeus. For the Vikings, lightning was produced by Thor as his hammer struck an anvil while riding his chariot across the clouds. In the East, early statues of Buddha show him carrying a thunderbolt with arrows at each end. Indian tribes in North America believed that lightning was due to the flashing feathers of a mystical bird whose flapping wings produced the sound of thunder." (http://thunder.msfc.nasa.gov/primer/). Personally we like the stories that say it is warriors clashing or angels bowling.

9.10 Jet Stream

We now know the Tropopause is not the same height (altitude) in all parts of the world. Warm air rising over the Equator can raise the Tropopause as high as ten miles. Conversely the cold air sinking over the Poles can lower the Tropopause to just over five miles. We now have the sudden, sharp temperature differences creating sharp pressure differences and sometimes splitting the Tropopause. The Tropopause is not a smooth, uniform layer of air but gigantic overlapping plates of air where extremely fast winds are generated. These winds are known as jet streams. The wind moves in squashed tubes thousands of miles long and hundreds of miles in width. These 'tubes' can have winds approaching 85 miles per hour compared to surface winds which rarely average over 25 miles per hour.

These Jet Streams blow mostly from the west to the east, which makes them an aid to airplane relative ground speeds. But this only works for travel west to east. Remember cyclonic winds in general are counterclockwise in contrast to the trade winds. As we said before, Jet Streams are a reaction to temperature differentials that are extreme and cause extreme wind speeds.

Since you as a beach walker can't measure the jet stream directly, and its speed, it isn't cheating to look at the TV or newspaper to see the patterns. Up to date folks will punch it up on their computer and that will help them guess the weather where they are.

9.11 Trade Winds

Trade winds mystified the early explorers in the Atlantic, until they figured it out. Trade winds have their direction determined by the Coriolis Effect.

The earth rotates from West to East. This changes the basic north/south flow of air over the earth. Winds in the northern hemisphere swerve to their right and winds in their southern hemisphere swerve to their left. This gives us mainly westerly and easterly prevailing winds.

The westerly and easterly winds come about because air is warmed at the Equator. It rises and blows, as wind, toward the Poles. At about 30 degrees north and south, air piles up and makes belts of high pressure. Some of this air cools, sinks, and blows as wind to the Equator. Because of the earth's rotation, these winds swerve off course and become northeasterly or southeasterly. These winds are known as 'trade winds.'

9.12 Storm Fronts

<u>Vilhelm Bjerknes</u> (1862-1951) is considered by many to be one of the founders of modern meteorology and weather forecasting. He coined the word fronts to describe the boundaries between warm and cold air masses.

The differences between the Polar air masses and the tropical air masses create depressions bringing rainy, stormy, unsettled weather also known as a storm front. The Polar air is moving toward the Equator and the tropical air is moving northward. The warm tropical air rising over the colder Polar air creates a drop in air pressure at the tip of the warm air which then becomes the center of the depression. Cold Polar air rushes in to replace the rising warm air and spirals into the center of the depression.

The edge of the advancing warm air is called a warm front. These warm fronts move relatively slowly about 10 to 20 miles per hour. The warm air is rising gradually over the colder air, forming clouds that may result in rain over a large area. Clouds form along a front. Since the cold air is moving faster than the warm air, they merge into a single rainy front. We call this an occluded front. Occluded fronts are marked by areas of rain, look for a low barometer reading. They result from a cold front overtaking a warm front.

An approaching depression will be marked by cirrus clouds, but as the warm front arrives, the clouds change to Stratus and heavy rain or snow falls. As the warm front passes, the air pressure stabilizes and temperatures rise. Hence, we have calmer, milder weather, but fog may develop.

The edge of advancing cold air is called a cold front. A cold front forces warm air to rise rapidly which brings gusts of wind with another pressure drop, but stable temperatures. Then the cumulus clouds appear with heavy rain, or snow, or hail and lightning. Once the cold front has passed, the temperature may drop, but the air pressure rises and clear skies return.

Cyclogenisis is the fancy word for the low pressure cells forming when cold and warm air interacts to form rotating weather systems.

9.13 Summation

That is our look at weather phenomena. Most of the discussion is in relation to the Oregon Coast, where we live. The details as to why weather happens are not as important as what may happen.

Generally speaking the most destructive thing you might encounter here is high winds and storm driven waves. Most of the time they happen together, but we are coastal, and the ocean is tricky and must be respected. Logs can fly like match sticks, as can occasionally rocks, when the ocean wants to show us a big wave. On a nice windy day it may just be pretty foam blowing around on the sand.

Air in motion is our weather. The earth rotates on a tilted axis. Positive or negative charges exist above the surface and negative charges exist at the surface. The sun heats our surface and surrounding atmosphere unevenly. All these factors are the heart of what we call weather phenomena.

10. THEN AND NOW

"Hey mom or dad, drive me to the school bus. It's raining outside". "Well now son, let me tell you, in ought six I had to travel umpteen miles in umpteen feet of snow. It was so cold you could spit and it froze before it hit. Besides that, when the old creek flooded we swam across with our books on our heads. What's your problem? A little rain never stopped me from getting to school".

Now, to get real about it, old dad needs to be careful about stories like that for the Oregon coast. In general, these stories of days gone by flourish in other parts of the country. However, here on the Oregon coast, the temperate weather has not changed much. There is not enough dense urban sprawl that would affect the climate. For example, in Portland (Oregon), the Willamette River has frozen over in days gone by. Never say never but with the heat generated by the modern urban area it would have to be even colder than it was in the past and stay cold longer to freeze the river. So far there is not enough urban development here in Lincoln City to change the climate radically. So what we get now as weather is pretty much the same as it always has been.

Cold weather is not expected on the coast and it is reflected in the building codes. Footings and pipes are not as deep as they would be in Northern Minnesota. Insulation in older homes was at a minimum. The new codes do attempt to conserve energy by requiring more protection. For example, there was a freeze along the coast in the 1970s. Houses froze, pipes burst, and there was a water crisis on hand because water was running out of all the vacation homes along the beach. No, the people were not dumb. There was no reason to expect this kind of hard freeze.

So, old dad can't use that old phrase on his son. When it does snow, it usually does not last very long. Those of you from the snow country really wonder what is the matter with these people. Two inches of snow and they talk about closing the schools.

Flooding is more likely to occur. Rain stories have a tendency to be true. Fortunately, the major flooding of the 60s and 90s seldom happen on the coast. Most cities and towns are expecting the worst these days since clear cut logging has increased the chance of runoff and of slides damming streams that will make matters worse. Urban growth with paving replacing wet lands and resulting in smaller estuaries, allows more run off and fewer places for the water to run to. Therefore we have more flooding.

10.1 Extreme Weather (?) Along the Coast

Looking at the record, weather extremes on the coast are floods, windstorms and snow in that order of frequency. The weather was severe enough to close the schools the winter of 1989. In the 1930s on October 21, 1934, December 31, 1938, and January of 1939, they used the word 'hurricane on the coast to label the storms. Most people remember the Columbus Day storm of 1962 which did its worst in the Willamette valley but had enough left to tear up things along the coast. In fact, during the middle 60s several storms did things like hurl logs through beach level rooms on at the Inn at Spanish Head. 1998 saw a repeat of this.

The 1920s newspapers show articles noting snow falling and freezing temperatures along the coast. The middle 1970s had another spate of windstorms. A 1980 article talks about 'the major snow storm.' Are these anomalies, never to be repeated, or just the frequent departures from normal climate? For the weather forecaster these happenings are why sophisticated monitoring equipment is so important to predict what may happen from what is happening.

Reviewing pioneer history of North Lincoln County Volumes I and II shows reports of cold weather, snow, and flooding, with occasional wind storms which kept the pioneers busy. However, when we look closer at the reports it's usually about the difficulty they had with weather in other parts of Oregon, as in the Columbia Gorge. The pioneers were more likely to talk about the gloomy periods of rain, drizzle that brought the short growing season (the green tomato year). As Lewis and Clark chronicled, the one continuous thread was rain, rain and more rain. The journal edited by Bernard DeVoto allowed us to review their stay and we felt it was a shame they hit a winter that was dreary to the extreme. Their journal entries of November 2, 1802 start off with sentences like "the fog so thick you could not see a man 500 steps off" or the November 5, 1805 entry; "It rained all the after part of last night. Rain continues this morning. Cloudy, foggy, and rainy. The only variation was high winds that made the river impassable."

However, when it did not rain, and the sun came out they felt just like we current beach folk feel. Their spirits soared and all was right with the world. Lucky for them, despite the problem of keeping their powder dry, they had elk hides to replace their clothing that literally rotted off.

Easter 1997 conditions were bad. March had arrived with 46 mph winds and 20 straight days of rain. According to the US Coast Guard SE winds reached over 64 miles per hour and the sea built to 20 foot swells. Precipitation moved from showers to thunderstorms and increased to hail for Easter Sunday.

Our friends to the north in Tillamook, and south along the Siletz River have seen flooding several years in a row. The weather record bears out the fact that the beach area around Lincoln City is temperate and usually does not anti up major weather problems. Long periods of rain are to be expected. This keeps our area beautifully green but it sure can get to you! Currently, with all the modern conveniences, a cold snap (as long as the power stays on) is generally not a big problem. Most people seem to wince about the occasional 'hot' spell when temperatures reach the upper 80s or lower 90s.

10.2 Are we affected by Weather?

What affects us? Any fool knows weather affects us. To keep focused on the problem, it is necessary to know that despite what you can see or feel, such as rain, sun, or snow, there are other effects that may be too subtle to pickup on until it is too late. Make no bones about it; weather can cause dangerous side effects on our human machine. We don't mean something obvious like a tree falling on our heads. Look at it this way, the humidity, barometric pressure, and wind chill need to be taken into account. In the sanitized city, air conditioners in the car, air conditioners in the house, air conditioners in the office with fixed windows keeps those folks isolated from weather's effects.

10.3 Weather Wise on the Beach

Now for a word about Heat exhaustion and Heat stroke. Heat exhaustion generally manifests itself when you feel dizzy, fatigued and get a headache because your body cannot endure the extremes in temperature. Heat stroke is by far more dangerous, and can lead to shock symptoms that if not treated, may lead to serious complications or death.

On the beach all of the factors can come into play on any given day. The Yoga devotees say listen to your body as you strike a pose or when you meditate. Our suggestion is understand how your body copes with extremes of weather and pay attention to the signals. Place a thermometer in your house with inside and outside temperatures. Put one in your car so you are aware of what is outside your climate controlled people mover. You will then have some idea what to expect when you must get out of your mobile cocoon. If the thermometer says its cold outside you should by now know enough to note the wind and know if you should stand in a protected area cover up appropriately to stop body heat loss and the possibility of frost bite on fingers, ears, or cheeks.

Most of us do not play in the ocean. The ones that do, like the surfers, take precautions. Exposure to the typical ocean temperatures will start a drop in core body temperature long before you realize it. Typical danger is a hot sunny beach in calm weather so you jump into the water. Then the old North wind picks up and wind-chill starts. On the best summer day at the coast, ocean temperatures are usually never very warm. Don't forget to refer to our previous discussion on hypothermia.

Despite the fact that we are not noted for a lot of days of sunshine, we do have a sun danger. It is possible to get sunburn. Not only is the sun affecting you, but also the glare off the water with or without wind or clouds can mean 'sunburn.' Sunburn is not related to high temperatures.

Those of you from other places with warm weather periods probably wish we had more of a problem with high temperatures here at the ocean. Since we usually do not get into the danger zone (above 90°F), if it happens you really need to be careful and follow the rules; drink fluids, slow down, stay in the shade if possible. The statistics at the local hospital probably don't show many customers with heat problems here on the beach.

Anyway, stay alert to the fact that your body will try to maintain the normal range of temperature if you give it half a chance. If you do not cooperate you will get hammered. Get used to reading the signs of weather changes and tide changes. A change in wind direction, speed, and temperature are warning signals. Your body really doesn't give you much warning before the system shuts down. Lucky for us, when it does shut down it is trying to save our lives.

The weather does affect our health, and it isn't easy to know if you are getting into the danger zone.

So be happy, and enjoy a great and variable climate, all because of that beautiful blue Pacific at our front door.

10.4 Summation

If you come from the city your climate controlled office, car, and home pretty well isolate you from the weather. When you join us on the beach you need a few of the old fashioned weather smarts. The whole idea is to get on the beach isn't it? On the central coast don't let past history and our temperate climate lull you into a false sense of well being. Some people are smart enough to know that when there is six feet of snow outside with a blizzard howling, that they need to stay out of the weather or dress appropriately. So when you visit the beach, do not forget the suggestion for being weather smart. Keep in mind that weather usually changes rapidly, so be prepared to leave the beach or change your mode of dress. You are more likely to get cold than too hot at the beach but either extreme is here. Watch the clouds, watch the waves, and layer your clothes. The extreme cold snaps or deep snows will seldom plague us. If you don't live on a flood plain you will occasionally be inconvenienced by detours around flooding.

From what we have been able to piece together, the weather in the past was pretty much as it is now. Mostly the weather seems to be rainy, winter starts in October, and possibly snow from December through February. Summers not too hot and mostly dry but do not bet on it. Late August through October is moderate weather which translates into our best weather.

The winds are usually here year round with direction related to the season. But the winter storm warnings need to be taken seriously since we can get buffeted by high winds. Wind can be seen as our atmosphere's reaction to two forces.

First is the rotation of our tilted sphere (earth) known as the Coriolis Effect. Second, the wind's force is a result of atmospheric pressure changes due to heat variations. Winds move from areas of high pressure to areas of low pressure.

There is nothing better than being in a storm at the beach or enjoying a sunny day in the surf, or just a plain old drizzly drippy day. As long as you heed our 'being weather smart' tips it will be fun. Do not forget to keep an eye out for the 'sneaker waves' and do not play on beached logs.

11. HOW TO GUESS THE WEATHER

When do we learn how to forecast the weather?

Well, this is that chapter! We probably should say "guessing" what the weather will be at a later time on another day. Probably the most fascinating thing about weather guessing along the coast is that you don't have to wait long before something else happens. What we all ought to be able to do is look around, and determine what is coming at us next in the form of weather.

11.1 Without Instruments

Let's start making your weather guess with only what our senses can tell us. See, feel (touch), hear, smell, and common sense are the criteria. This is the way we will be most of our lives since we don't walk around very often with portable weather stuff.

Don't forget the old sniff test. Some people say they can smell rain. Who's to argue with them? Lightning strikes ionize the air as does any electrical arc and that we can smell. On the coast we have noticed you do not often get that salty ocean whiff so common on the east coast. Occasionally you may smell that something has died but that is usually the sailor jacks or common jelly fish relative of the Portuguese man of war.

Hearing on the beach is important. The ocean sounds give clues about offshore winds before they disturb the land.

The visual clue is the best thing we have. Whitecap activity indicates wind speed, and the direction indicates where your weather is coming from. What you see may suggest how much rain is falling. Since most of our rain comes in over the ocean, sharpen your cloud knowledge to actually see the rain coming.

Remember you feel groggy or more tired in high humidity. Feelings are important in forecasting short-term changes in the weather.

With the miniaturization of weather equipment, portability has arrived. Can you see people with little anemometers on their beanies? There are hand held anemometers with wind speed and wind chill, as well as a watch size barometer and pocket hygrometers.

Let's get in the habit of looking around when you step out of the house. Start by looking at the clouds and checking how far you can see, provided your view of the sky is not hidden by the neighbors or a hill. Next, notice which direction the wind is coming from and what the trees and bushes are doing. Write down your observations. Don't forget the old wet finger routine when the wind is so calm that nothing is stirring. Believe it or not, it works. This will tell you the direction from which a little breeze is coming. Due to the evaporation of moisture from your skin, the cold side of your finger is the windward side. Your neighbors may think you are a little eccentric, seeing you sticking a finger in your mouth and holding it aloft like a lightning rod.

A good cold temperature indication is when you can see your breath. It is your own personal fog cloud of ice crystals. Most of us know that the temperature is down when our breath creates fog as we exhale. If you see fog outside and it evaporates (dissipates) quickly, it heralds fair weather.

Do notice the birds or other furry friends as they do their thing. They will provide clues to the weather. They will help you forecast if the temperature is going to be extremely high or low and if it is going to rain or get stormy.

Once those observations are made, you will begin to get weather clues as to what the weather may do, without using instruments. You will find yourself in good company.

Contractors, loggers, and the like, must plan their activities around the weather. They do not carry sophisticated weather instruments. They use the simple techniques we are providing you.

11.1.1 The Media

If nothing else, start your day by listening to the TV weather guys or better yet, the local radio people who at least live near you. Forecasts can not be site specific, so don't be mad when the radio guy in the town 24 miles away says it's nice and sunny here today when you are ducking a squall that just rolled in.

If you are the type to enjoy non-technical information and unlimited amounts of weather trivia, read the daily forecast for your section of the country in your favorite newspaper. You will soon get a feel for how accurate these forecasts are. They seem to be quite accurate especially with major weather problems like floods or snow or ice.

The Old Farmer's Almanac is fun to read and follow the projections which are made a year in advance. Be sure to put a string through the hole in the corner so you can hang it in your favorite reading spot. If you don't use that hole they may stop putting it in each issue of the Farmer's Almanac.

11.1.2 Wind

The first thing you may notice on the coast, after you stare in awe at the mighty Pacific, is the wind, or lack thereof. You can't see wind but you can feel it and see what it does to our belongings and us.

If a wind storm really blows in it will tear-up our houses and fences. The wind around our beach environment generally comes from either North or South. If you stand with your back to the wind the low pressure area is always on your left in our northern hemisphere. With this information you will know the location of the low.

Remember, the wind blows from the high pressure to the area of low pressure. Based on the wind speed and direction, you can begin to guess if the low will get to your section of the coast. This brings in the stormy weather.

When the old wind is from the northwest, it could be pushing weather on shore, but at least it will blow out the fog and make those clouds move on before they get you wet. A north gusty wind generally means a cold front can come in during the winter, but during the summer it keeps us cool. During the day, when the northerly winds are blowing, you need to keep an eye out for a swing to the south which could indicate incoming rain.

The headlands modify the southerly winds in winter and the northerly winds in summer. Our Central Oregon coast is rather straight, running North and South, but individual areas can slant East or West. There have been times in Lincoln City we have avoided damaging high winds while the headlands and other areas experience severe wind damage. These headlands may be a basalt volcanic plug, not worn down by the ocean, or sandstone bluffs that rise above the beach.

Those of you that have come from the East coast shoreline or from some island paradise expect a breeze blowing on shore or off shore. Usually coastal winds are land-sea breezes. During the day, the land heats up faster than the sea and the air rising over the land is replaced by cooler ocean air. This is an on shore breeze. At night, the land cools faster than the ocean, so the air moves off shore, toward the ocean. This is not the case at our central coast beach since the blue Pacific generally stays approximately the same temperature (cool), which is cooler than the land. Of course, our land never becomes extremely warm either. Therefore, the usual on/off shore flow doesn't happen here as it does on the east coast. What we get are seasonal north or south prevailing winds. These winds are generally from the north in the summer and from the south in the winter. Northerly winds usually mean clear while winds from the south indicate rain.

The next obvious question is; how fast is the wind blowing?

Smoke that drifts shows wind direction. Water ripples, but no sea foam, indicates 1-3 mph light breeze.

Beach grass swinging to and fro indicates 8-12 mph gentle breeze.

When the sand is moving fast enough around your ankles to level out the beach, and your beach umbrella does not stay put, and there are waves with numerous white caps as far as you can see, these indicate 13-18 mph moderate breeze.

Northwesterly winds usually mean fair weather. When the wind starts to shift to the SE you probably will get rain. If the winds continue to rotate from the south to the southwest you could get clearing.

Do not forget storms can come from any point on the compass. These are only guidelines.

Large waves and trees swinging to and fro indicate 19-24 mph "fresh" breeze.

Larger waves with whitecaps everywhere and more sea foam spray indicate a 25-31 mph strong breeze.

If you look out at the breaking waves and the sea foam takes off with streaks of foam, rain is horizontal, and then it is blowing about 32-38 mph which is a moderate gale.

When you are about to launch your boat, look for weather indications, before you launch. The Coast Guard is good at posting flags and sometimes light signals to indicate expected weather.

The red pennants are most prominent.

1.   One Red is small craft warning with winds up to 33

knots

2. Two Red is Gale warning with winds 34 to 47 knots
3. One Red Square with a white center is a storm with winds above 47 knots
4. Two Squares is Hurricane with winds above 64 knots

At night lights (and sometimes globes) are used to indicate the above conditions.

1. Red above white means winds to 33 knots
2. White above Red; winds 34 to 47 knots
3. Two Red; winds above 47 knots
4. Red above White followed by another Red; Hurricane with winds above 64 knots

Don't forget the faster the wind blows the quicker the weather changes. Also keep in mind the direction clues.

As we have said before, the wind comes from the south in winter and generally has some kind of storm attached. The wind can also come roaring in from the north fast enough to level beach sand. This should drive the clouds away to make for a sunny day on the beach. The wind will still make it feel rather chilly.

11.1.3 Clouds

When it comes to clouds, the coast is a great place to be a cloud watcher. Read dark fog and clouds on the horizon (over the ocean) as rain. But how many minutes will it take to come ashore? This will take repeated observations including estimates of the wind speed.

The next best thing to give you a clue for a weather guess is the clouds or lack thereof. Now that you know a few cloud names and shapes, what do they tell you? Those high wispy cirrus mares' tails are usually the leading edge of a front; especially on our beach area. If they are in the east in the morning, you may get a weather change before the day is out. If they are in the west, you have time to enjoy some clear weather before anything happens.

Do not forget the obvious - no clouds equals no precipitation. There are many identified cloud formations. Here are some of the most obvious ones that give you weather clues.

Anvil (flat topped) cumulonimbus can rise up to the Cirrocumulus height. If you have forgot what an anvil looks like visit your local blacksmith. The anvil is formed when the cloud can rise no farther, so it flattens as the above wind shears off the moisture. Anvil cumulonimbus and Cirrocumulus appear in combinations and variations. Large blobs hanging down from the cumulonimbus like a giant udder are called Mamma. These indicate strong winds and heavy rain.

Some areas (not our coast) have lenticular clouds. These are altocumuli lens-shaped clouds around mountain peaks due to a wave in the air stream passing over the peak.

Remember the old nimbus clouds, they really look full of water and most likely will dump somewhere. The squall fronts off the coast really need to be watched if you are worried about getting wet. They are just as likely to miss you with the west blowing them to Tillamook. At best you can bet it is only a squall and it will clear up soon. Don't forget, sometimes the squall fronts line up out there, one behind the other.

So what do you do with that sighting? Remember! Clouds are fog; fog is water vapor; rain is heavy fog. If you see clouds, water vapor is accumulating in your air. If you have wind, this water vapor may be passing through. You may not get it, but it is possible someone east of you will get the moisture.

Hazy clouds? possible storm within hours

Cauliflower like tops? Imminent thunderstorm

Rolling dark cloud? Approach of bad weather

Fleecy white clouds? Indicates good weather

High wispy clouds (cirrus)? Indicates a front is coming

Or we can use proper names if you remember what they look like

CUMULUS - clearing weather (the rain should go away)

STRATUS -may thicken into nimbostratus with rain

CIRRUS - behind or leading a front to fair weather

NIMBUS - precipitation

NIMBOSTRATUS - rain or snow

CUMULUSCONGESTUS - rain or perhaps just showers

CUMULONIMBUS - severe weather (tornado watch)

The nimbus clouds usually form the cloud deck that makes the beach a little dreary looking. It is a good clue as to weather (pun intended) it will rain or not. The clouds are lumpy on the bottom and they really look full of rain or not so full. If they look full and do not seem to be thinning our bet is on rain.

Cirrocumulus is higher; above 18,500 feet. This is sometimes called a 'mackerel sky' and has the pattern that the wind makes on our sandy beaches.

Stratocumulus clouds are lower; up to 6,000 feet.

Obviously, no clouds in sight means it will be a while before you get rain - no clouds no rain. Low, fat, and heavy clouds means rain is imminent.

There is no accurate way for an amateur weather person to determine the extent of cloud cover by just looking. Satellites do this job for us. When you have a cloud deck you get clues as to how thick it is by how bright it is outside. A thinning cloud deck and obvious brighter sky is a clue it is going to break up.

The sky has been a weather forecast source from early times: According to the King James version of The Bible, Matthew 16:2 and 3, Red sky in the morning, Sailor take

Warning (foul weather), Red sky at Night, Sailor's Delight (fair weather). Most of our weather systems on the coast approach from the West, so red sky in the morning means the sun rising in the east is reflecting off cirrus clouds moving in from the west. These thin and wispy clouds are composed mostly of ice crystals. Cirrus clouds typically appear at the beginning or end of a storm front and are therefore, early signs of a storm system.

11.1.4 Temperature

When it comes to guessing the temperature outside be prepared to get fooled. One of the coast's favorite ways to fool people is when the sun comes out they think it is warm out. Not to be, since there usually is a brisk cool breeze.

If you see your breath, or see frost on the grass, you can bet it is cold. If you <u>feel</u> warm it probably is warm. Occasionally we get clues when we have extremes in weather. A side bet is that if you can see ice in the puddles of the pot holes on our side streets you can bet it is freezing. It is usually that cold only for a few days and it does not happen every winter here on the beach.

Here are some other clues about extremes in temperature. It is hot if the birds are in the shade and quiet, some of them holding their wings away from their bodies. Do not be fooled by the cormorants that do not have waterproof wings. They hold their wings out to dry them, and not because of the heat.

The birds are cold if they are actively chasing around for food to fuel their bodies since the colder it is the more food they need. When they sit around all fluffy to twice their size, it is to increase the warm air space next to their bodies. People get worried about one legged seagulls in their area. If you are patient you see it swap legs. The gulls get cold feet and will hunker down and cover up both feet or stand on one leg to warm the other.

Birds tend to perch more when a storm is approaching. The key is barometric pressure. Lower air pressure lowers density and thus less support for flight. So why strain, just perch as the storm approaches.

We also can't help using our outside cats as a guide to temperature. If they are in the sunny spots, stretched out, it is fairly warm. If they are in a sunny spot all tucked in, it is pretty cool. If you do not see them around you can bet they are tucked away in some cubbyhole in an attempt to keep warm and out of the wind.

11.1.5 Precipitation

When it is raining sideways most any fool can guess, Man, it's really raining. As weather watchers, we can come up with some general terms like drizzle, raining buckets, etc. The typical tipping buckets in rain gauges are not sensitive enough to react to drizzle, but an all day drizzle can build up to 3 or 4 tenths of an inch of precipitation.

Here is the closest guide we can find.

Drizzle:	Drops with a diameter less than. 02 inch, falling close together. They appear to float in air currents, but unlike fog, do fall to the ground.
Light Drizzle:	Visibility more than 5/8 mile.
Moderate Drizzle:	Visibility from 5/16 to 5/8 mile.
Heavy Drizzle:	Visibility less than 5/16 mile.
Rain:	Drops larger than .02 inch or smaller drops that are widely separated.
Light Rain:	0.1 inch or less in an hour. Individual drops easily seen.
Moderate Rain:	.11 to .30 inches per hour. These drops are not clearly seen.
Heavy Rain:	More than .30 inches per hour. It seems to fall in sheets, reducing visibility.

11.2 With Instruments

Now that you have some idea how to look around and get good clues to the weather changes it is time to make you a weather guru, based on instruments. The only thing the instruments will do for you is to give you additional information and more accurate data.

Despite the fact that you have a weather station it is still a good idea to do your visual check of conditions. This really becomes a part of your prognostication skills. In fact, it becomes important to make estimates of wind speed, direction, temperature, and barometer, to know if your equipment needs checking. Instruments are not fool proof. Nothing gives a false reading quicker than seagull droppings in your rain dipper funnel. Corroded electrical connections are another popular cause of a misreading in this salty air.

It is assumed that you are going to log the data on your own charts. If you have one of the digital logging stations, select the time interval that serves your needs. Generally you do not need readings 15-30 minutes apart unless you are tracking a storm. When that happens, the shorter the time interval the better to tip you off to arrival time, and better yet when it will blow by. Another problem is that very frequent readings on a daily basis may overload your computer's memory without providing any critical information for weather guesses.

We suggest you select a time of day convenient to log the weather and clear your instruments. This becomes very important when you are looking for patterns in your data or when you are comparing your data to the weather guys or a friend. An example is the Roads End Weather Watcher. He has to keep two sets of data. One is for his station which is every 24 hours from approximately 6:30 a.m. to the same time the next day. For Roads End Weather Watcher's report to KATU the time cycle is from 4:00 PM to the same time the next day. These reports have revealed a pattern showing this is a time when the beach weather can shift. Sometimes the weather shifts for the better if the barometer is indicating an average low and is turning upward (higher pressure). When the low is very low (below 29.50) indicating a further drop, you can bet the weather will be taking a turn for the worse. So do not get upset when weather people say one thing and you have another. Do not hesitate to call the TV station and find out what their 24 hour cycle is. You most likely will find out they will respond to you in a positive way and if you are lucky you may be able to be one of their weather watchers. TV station people know that visual observation is very important in conjunction with their sophisticated equipment.

Regardless of what kind of system you setup, keep your sense of humor and adventure. In the beginning you may be more wrong than right, and as you get good at guessing, you can forecast further into the future. Then what happens? You start being wrong again. However, take our word for it; you will get good enough for what we need in our temperate coast.

By the way don't forget to check other sources of weather information. Start your day by listening to the TV weather or better yet, the local radio people who at least live near you.

Forecasts can not be site specific, so don't be mad when the radio station in the town ten miles away says "it's nice and sunny here today" when you are ducking a rain squall that just rolled in. Remember that on our Central Oregon coast the TV stations are 100 miles inland.

If you really feel the need to pin down weather forecasts, there are emergency phone numbers that give road conditions and specific problems. It takes a while but the audio tape will eventually get to your specific area if there is a problem. Examples are ODOT (Oregon Department of Transportation) phone numbers in your telephone book, telephone numbers in the weather section of the newspaper, your local travel service (like AAA), NOAA, radio stations, and similar sources. Do not forget the web sites for other agencies.

Back to those instruments you spent hard earned money on. Go ahead and take your daily readings. The basics are barometric pressure, wind direction, and speed, and a place for a good old fashioned 360 degree scan of the sky.

Once you have some historical data these concepts will begin to reveal themselves. Read over your charts with an aim to forecasting weather. Remember they are as site specific as possible. However, these charts do use common principles that work wherever you are in the world.

11.2.1 Wind

Start by noting the wind direction. Keep in mind that seasonally the wind comes from the north in the summer and the south in the winter. Since this is not rocket science there are no absolutes.

Once again let's remember what our winds can tell us. If the wind is from the north it blows the clouds away and generally it means clearing. If the wind is from the south it means rain. Do not get mad at us when you find that it can rain with the wind from any direction. Our winds are seldom directly from the East or West. If they are, it means the wind is changing and so is your weather. In some cases changes are for the worse and in others for the better. If it is a brisk wind, how fast is brisk? It is faster then no wind and the weather change will be quicker.

Steady winds, here on the coast, of about 10 miles per hour will cause weather changes but the 20-30 mph winds really shorten the time for the weather change to occur.

11.2.2 Air Pressure

The barometric pressure is your next best clue. In fact it is the one piece of data that looks further into the crystal ball than do the others. Think of 30.00 inches as the baseline. If you have an old fashioned aneroid barometer, give it a nice little tap with your finger, and notice the direction the needle moves. This movement is in relation to the manually adjusted needle you set over the aneroid needle the last time you took readings. This will tell you whether the air pressure is going up or down over time. Some instruments provide a simple forecast (sun or rain) using icons (picture of sun or of clouds and rain) while others may display an arrow pointing up, down, or horizontal. The horizontal arrow indicates a steady barometric pressure (no change).

If the pressure goes up, it usually forecasts clearing weather. If it goes down to 29.80 or lower, it usually indicates stormy weather. When it gets to 29.50 or less, it usually means serious winds and rain storms for the coast. This is stormy weather and you need more frequent data points. On the digital stations this is the time for reading the short data intervals or for periodic taps on your aneroid barometer to clue you into the seriousness of the change.

Rule of thumb, the faster the barometer falls the quicker you better batten down the hatches. Our experience here on the beach has been that a <u>fast</u> rising barometer has indicated good weather arriving soon. A steady barometer does not tell you what the change will be, but alerts you to a coming change. You need to watch closely for when the change starts.

Barometric pressure movements on either side of the baseline indicate the same thing as major swings. Our data has shown that a high barometric pressure of 32.00 that drops to 30.00 can predict storms. Likewise, a low pressure of 29.60 rising to 29.90 can mean clearing and secession of rain storms.

You should be able to improve your forecasting by noting how many ups or downs there are before something happens. These indicate the change point that seems to be pretty consistent at our stations. No matter how you go about it, we bet you will come up with your own clues.

Stronger winds mean a lower barometer or a <u>fast</u> rising barometer, 30.00 inches or higher mean better weather. Watch the direction of change. Low barometer going up could be better weather. High barometer going down could be storm coming.

11.2.3 Temperature

Temperature readings are more difficult to use for forecasting. If it is warm in the morning you can bet on the day being warm. As usual the rest of your gear should also be indicating fair weather. Barometric pressure ought to be high, the cloud cover either missing or light, and the wind blowing in a direction to not bring in cold air.

Low temperatures can be forecast if the temperature at sundown is close to freezing. Lucky for us we talk about 30 degrees Fahrenheit as cold. None of the minus 20 degree stuff for us on the beach.

Another thing to watch for is quicker than normal rises or drops in temperature. As day comes into focus, the temperature usually rises here on the beach as the day progresses, even in winter. After sunset the opposite generally happens. The high temperatures are usually in the middle of the afternoon depending on wind speeds and direction. The colder northerly winds keep the temperatures down. Occasionally, calm sets in and a spike in temperature will occur at sunset.

These observations of temperature will give you the ability to really guess when a warm or cold front is arriving or has moved through. As mentioned before, the temperatures will rise or drop as the front moves through.

How do the professionals forecast temperature highs and lows for days in advance? Good question! Now we come to the poor answer. They do it with computer data profiles with lots of historical data. In other words, it is not a deep scientific analysis, but history being counted upon to predict the future. History does a pretty good job. Remember the 'experts' are seeing fronts moving in from far off shore and know how fast they are moving and the air temperatures they are generating. None of this is happening for the first time, so it is just like we tell you with your home system the more history you compile, the more accurate you will become. Do not get too serious, since a complete miss is almost as much fun as being 100% accurate.

11.2.4 Humidity & Dew point

The last items to consider are Humidity and Dew point. Measuring humidity in the eighteenth century involved a hygrometer of paper disks on one end of a fulcrum with a needle and scale on the other end. As the paper absorbed moisture the needle moved up the scale.

Generally speaking, the humidity reading gives you a long range indication of weather change. On a clear day if the humidity goes up this indicates more moisture in the air and clouds may be forming. The humidity follows along with the barometer as a precipitation clue. If the humidity goes up, and the barometer goes down, we have a better chance of getting precipitation. The reverse is also true. No matter what, if there are no clouds, there will be no rain.

During cloudy weather, a lowering of humidity indicates clearing (fewer clouds). These concepts are based on reading the outside humidity although some instruments are only indicating inside humidity. Be aware of which instrument you have.

The more complete systems will give you dew point as well as temperature. The dew point is the temperature at which the water vapor in the atmosphere becomes liquid. Those of you that have the weather systems that give you the dew point data can add a few more factors to your weather guess.

The bigger the difference between the air temperature and dew point temperature, the dryer the air and the less chance for condensation. Condensation is needed to get some form of precipitation. When the temperature and dew point temperature are the same, fog or clouds will form. Keep in mind that when the Valley (interior) heats up, we are likely to get fog on the beach, especially if our northerly winds take a rest. If there is a heavy frost or dew early in the morning (or late in the evening), that is a good indication of good weather for the next 12 hours.

11.3 Summation

Keep in mind that on the coast the ocean is the most important weather control factor. The water gains and loses heat more slowly than land. Therefore, our weather is relatively bland with few extremes of high or low. If you are lucky some winter a cold north wind instead of the usual warmer south wind, you'll see spin drifts made of ice crystals.

This chapter is a long one. Don't expect to memorize all this material but start focusing on key elements in the chapter. The real purpose of this text is to help you become weather wise with only a practiced eye to guide you. We will say it again; if you leave your climate controlled house and venture out, you can be adversely affected by the weather. In some cases it is life or death when you misjudge what might happen next to your weather.

Our suggestion is to learn a few major cloud formations and get some clues as to wind direction and speed. Keep in mind the season of the year and you will find yourself making choices of dress or direction of travel based on what old Mother Nature will do next.

If you decide to be more organized about your weather guessing, keep in mind the following ideas. Based on your charts and visual observations you will have an historical data package which will improve your weather guesses dramatically. You will begin to see trends around your station, because as we have stated, there are micro climates, especially in our stretched out coastal area. Its fun to hear a local DJ talk about rain/hail or whatever, while folks several miles away are in the opposite type of weather. More dramatically are the differences between our coastal strip and only a few miles inland, especially our famous Devils Lake.

The hope is that we have convinced you of the need to be aware of weather forecasting and the fun you will have collecting data with weather gear. Do not forget the more sophisticated your gear is the more likely it is to go belly up during that storm. Rain gauges generally are bullet proof but eventually may wear out. Most of the gear has batteries so get on a schedule of replacing the batteries before they go out.

All in all, the more data available to you and the longer you have been recording, the better your guesses will be. A word of warning, do not get caught in the trap of helping some one plan an outdoor activity that needs clear weather. As soon as you predict clear weather, it will rain.

The most fun from recording data is keeping your memory straight. Somebody always pops up with remembering when it snowed on New Year's Eve. It did snow sometime but you will be able to pinpoint the date by going back to your log. Actually, on the coast it would be a very light snow, but let's face it, if the ground is covered it's a big snow.

Start collecting data from wherever and begin enjoying the ability to plan ahead for dress, trips, and outdoor activities. You can reasonably predict what the weather will be. Also do not be afraid to include your intuitive sense as well as the behavior of natures furry and feathered friends to help with your forecasting

It is never too late to be a weather watcher so get some gear and start recording data.

12. ODDS AND ENDS

12.1 The Sun

Let's start with the sun itself. The sun has a diameter 109 times that of earth 864,000 miles. The two have existed together some four and a half billion years. The sun is located a mere 93,000,000 miles from us. That is far enough away to allow earth to have water in liquid, solid, and gaseous states. Life depends on water.

The sun is what makes weather, with unequal heating aided by earth's rotation. Our only other heat sources are molten rock in the earth's core and radioactivity. Geysers and volcanic eruptions spew water and enough ash to darken the sky, but the impact is minor compared to the sun's daily work. Since 1938 the theory is that the sun is nuclear fusion. A fusion reaction is when atoms fuse together, creating other kinds of atoms, and giving off energy. This is in contrast to our managed fission reactions for power plants in which we split atoms to produce energy. Now we are back to unequal heating. The stronger the contrast between hot and cold, the stronger the storm will be.

Land and water absorb and release heat quite differently. Water bodies like the ocean, are reluctant to absorb or release heat compared to land. Try it. Walk the beach on a hot sunny day. The sand is <u>hot</u> but the ocean is <u>cool</u>. Try this again at midnight on a clear evening it appears the temperatures have reversed. Increasing our perspective, it becomes clearer why tropical waters remain warm and frigid polar water is just cold.

Next, there are what we call solar flares and the big boys call solar prominence which are huge eruptions of solar gas soar off into space and are best seen when there is an eclipse. The moon blocks out the sun, while leaving a corona of gases visible around the edges of the moon's shadow. Don't dare look directly at the eclipse or sun to see those flares. There are ways to use a shoe box to cast a pinhole view of the sun on the inside of the box without looking at the sun.

That big old sun keeps shining even though it may be hidden by the cloud cover. As a kid (which was many years ago) not everybody flew in airplanes trains and buses were more likely the mode of long distance travel, which do not afford much of a sky view. Anyway, not until we got into an airplane that could fly above the ceiling (that's pilot talk for clouds), did we realize the old sun was still there shining as brightly as ever.

Let's focus on the weather aspects. You can get another book that goes into the sun's origin, its atomic power plant, etc. You will not be disappointed. A little high tech astronomy stuff is good to know but even the experts are not sure what that has to do with the weather. For us amateur weather guys that live on the beach, all we need to know is the sun makes the world livable. Let's hope it is in no hurry to do what stars do; burn out.

12.2 Solar Flares

The solar flares occur in conjunction with those strange things called sunspots. As a result, massive storms of cosmic rays and X rays stream toward earth and do things to us that even the weather experts can not quantify.

The CB radio users were one of the first groups to directly experience the effects of these solar storms. During peak X-ray, cosmic and magnetic storms the feeble five watt transmitters would have so much background noise even the four letter words got drowned out. The next minute, on skip, you were talking to some guy in Clint, Texas as if he was next door. All types of communication are affected, including broadcast radio, television, computers, and satellite hookups.

Another visible effect of this type of sun activity is the aurora. Rarely have beach dwellers seen it, maybe on a clear cold winter night. Normally the aurora is a show saved for our far northern friends. But we all can experience the wonderful crepuscular rays.

Crepuscular is the name for the sun rays we see peaking through holes in the clouds that remind us of the Japanese Rising Sun symbol.

12.3 The Almanac

When it comes to long range forecasting the <u>Old Farmers Almanac</u> has got to be the champ. Started in 1792, the almanac has developed a system of forecasts almost two years in advance, when you consider how soon they need to start their forecast in time to complete publication of the book which is October of each year. Don't bet your life's savings on the daily forecast, but if you track it in your local media, you soon get a feel for how close it is or how to extrapolate some general forecasts that are quite accurate.

Bet you did not know that it cost approximately $43,000 to drill that hole in the upper corner of the book. If you noticed it, we will bet even money that you do not know anyone using the hole. In the good old days (not as far as we are concerned) you had an additional building on your property situated (hopefully) a ways off downhill away from prevailing winds and underground water sources. In this other essential structure a feature would be a copy of the Almanac hanging by a cord looped through the hole and hanging on a nail close by. Our guess is that last years almanac got recycled in this very same structure. Enough said!

12.4 The Beaufort Scale

Sir Francis Beaufort, a British Navy hydrographer developed the scale in 1805. It measures the wind's effect which then tells you the approximate speed of the wind. He noted a more precise and concise terminology was needed to describe winds at sea.

The Beaufort Scale:

Smoke goes straight up	Calm	less than 1 mph
Smoke drifts	light air	1-3 mph
Leaves rustle	light breeze	4-6 mph
Light flags are extended	gentle breeze	7-10 mph
Dust & small branches blowing	moderate	11-16 mph
Small trees sway; crested waves	fresh breeze	17-21 mph
Moves large branches	strong breeze	22-27 mph
Whole trees in motion	moderate gale	28-33 mph
Walking is impeded	fresh gale	34-40 mph
Structural damage	strong gale	41-47 mph
Uprooting trees	whole gale	48-55 mph
Lots of damage	storm	56-65 mph
Excessive damage	Hurricane Force	65+ mph

12.5 Regional Winds

Regional winds are named by the specific geographic areas in which they occur. Another descriptive term is local winds. The simplest is the land sea breeze along our coast. When the land is heated, the air rises and the onshore flow begins. Likewise, when the land is cooling faster than the ocean, the wind blows offshore. The regional winds name's mostly relate to the effects of these winds. We don't appear to have any names for the winds on the central coast but, for your information, here is a list of six that might be familiar.

Chinook- This is a warm wind down the East slope of the Rocky Mountains that clears the winter snow for hundreds of miles along the high plains. An example is Eastern Montana.

Haboob- These are severe dust storms occurring mostly in the Sudan. They typically are walls of dust that can be thousands of feet high.

Santa-Ana- This warm wind descends from the mountains into Southern California. Like the Chinook, it is a drying wind and therefore can be dangerous for already dry hillsides if any fire is present.

Sirocco- Again we are talking about a warm wind but it is in the Mediterranean. It sweeps north from the hot Sahara.

Katabatic- These are mountain winds mostly in Norway. On clear nights, air in contact with the cold snow on the mountains is cooled rapidly in the evening. This sinking air drops to valleys and makes crop damaging frosts.

Monsoon- In June, winds from the Indian Ocean and the Bay of Bengal blow on shore, full of moisture. They rise as they blow over high ground and cool, forming clouds and torrential rains begin to fall. In the fall and winter these winds blow offshore to the warmer oceans and make the land even more barren and arid.

Air pressure creates the wind. This includes the winds traveling hundreds of miles across the oceans and land. These winds may be light breezes. Since the air pressure at any location rarely remains steady for long, weather is in constant change as winds ebb and flow.

12.6 Hydrologic Cycle

Here is another of the exciting dynamics going on in our weather world. Our favorite resource, water (H_2O), gets recycled on a regular basis. The 'water cycle' is instrumental in cloud formation and resulting precipitation. Water gets into the atmosphere through evaporation from seas, lakes, rivers, ponds, wetlands, and our ocean. The warm air is moisture laden, so when it cools, the water vapor condenses, forming clouds, and inevitably falling back to earth as rain, hail, snow, sleet etc. When it rains, if we do not suck-up the water and do something else with it, it runs down hill to eventually find its way back into the ocean.

The minerals needed in the ocean are carried back from the land through runoff. Remember also that rain started around some 'dirt' particle in the atmosphere.

This is the hydrologic cycle and if you are into scientific terms the cooling of the air is caused by air masses moving and can be 'adiabatic' or simply 'by contact.' Air cools at about 5.5 F° for each 1,000 feet of elevation gain and warms at approximately the same rate for loss of elevation. Whether from crossing mountains or just movement in the atmosphere, this temperature change is called Adiabatic.

12.7 Seasons

Earth's orbit around the sun is at an angle of 23.4 degrees which accounts for the seasons. OK! We are spinning one revolution every 24 hours around our axis and orbiting the sun once every 365 days. Since the plane of our orbit in relation to the sun is constant we experience three distinct angles to the sun in our 365 day rotation. In summer, we in the Northern Hemisphere, note the North Pole is angled toward the sun, so our days are longer and warmer than in winter. This reverses in winter so the sun is on the South Pole more. In spring and fall, the poles are relatively equidistant from the sun. The Solstice and Equinox are calendar events representative of these events. Equi means equal so Equinox is when the sun is in line with the earth on its axes. Solstice means angled toward the sun at the 23.4 degree tilt.

Obviously, we tilt toward the sun for summer solstice. The North Pole is tilted directly toward the sun during the summer solstice; around June 21st. This is our longest day here in the Northern Hemisphere. The sun is actually directly over the Tropic of Cancer (north of Hawaii). This is the northernmost spot that the sun can be directly overhead. This imaginary line runs through the United States, approximately at Key West Florida. Therefore, in Hawaii in May, the sun is moving northward toward the summer solstice.

At the spring (March 21st) and fall (September 21st) equinoxes the sun is directly above the equator. It rises almost exactly due east and sets in the west. Day and night are almost equally long all over the earth.

Spring and fall equinoxes are repeats of the sun's effects. The difference is that as we come to the fall equinox, the North Pole is farther and farther away from the sun (at earth's 23.4 degree tilt) and we go back into winter. At this time the sun is directly above the Tropic of Capricorn which passes through South Africa, Brazil latitudes. This is the lowest point for the sun in our sky; the Northern Hemisphere.

Wow! But, we thought you should know.

12.8 The Water Years

Rain tracking is a popular sport. Precipitation is usually averaged monthly and on a calendar year basis. However, the term "water year" is now being used consistently. This 'year' starts October 1st and ends September 30th of the following year. This gives a more accurate picture in areas where there is a winter maximum and a summer minimum precipitation such as on our Oregon coast. The following table is applicable to the Central Oregon Coast.

WATER YEAR	TOTAL Inches	CALENDAR YEAR
1985	60.5	73.9
1986	75.1	62.7
1987	87	72.7
1988	82.78	89.5
1989	87.55	85.4
1990	63.71	75.64
1991	106.7	83.96
1992	58.88	93.9
1993	53.7	54.98
1994	45.65	46.9
1995	78.1	68
1996	74.18	73.75
1997	117.44	96.05
1998	78.24	90.2
1999	95.09	98.5
2000	45.90	57.35
2001	79.71	63.19
2002	67.13	75.08
2003	67.5	75.04

12.9 Tsunami

What is a tsunami and why talk about it in a weather book? A tsunami is a wave generated by forces under the ocean aside from wind and tides. There is no particular connection with tsunamis and weather but if you are going to be in a beach community you better listen up.

The ocean is never still and usually has waves, tides, and currents. A tsunami generated wave can be devastating when it reaches the shoreline but vessels at sea may be almost unaffected. The tsunami is not generated by the force or the existence of wind.

Americans earlier called it a tidal wave, thinking it had to do with some kind of tide, such as the Tidal boar on the Bay of Fundy. The water roars in and fills the bay and later moves out leaving the bottom exposed for miles. That is how a large tsunami wave behaves. Therein lays the trouble. When the water recedes, (we will discuss why later) the bottom is exposed. This may be the first time coastal residents see the exposed bottom. So guess what, they all run out on the flats to pick up stuff. We now know what happens next. The water returns, so don't go out. That exposed ocean bottom may be your only warning that a tsunami is on the way. The water will return in a series of waves.

The Japanese call the wave a tsunami which means harbor wave. It has happened to them many times in their recorded history. On our coast, the last time was over 300 years ago (January 26th at 9 p.m. 1700). The core samples on our coast suggest a tsunami comes along about every 300-500 years.

What causes them? Some kind of under ocean movement must take place to move huge volumes of water either up or down. The most obvious thing would be an under ocean earthquake. However, it needs to be a particular type of earthquake with special plate movement. If the plates slide parallel to each other a tsunami may not be generated. But if the subduction zone quake takes place the movement is likely to cause a tsunami.

The beach where we walk and live is on the American plate. The Juan de Fuca plate slides under the American plate and causes upward deformation of the American plate. I'll bet you didn't know the house you live in are being pushed up slowly but surely. Currently the two plates are locked together. When the plate slips the American plate slides west and drops six or more feet. The part of the American plate that reaches off shore as well as the Juan de Fuca plate also drop out from under the water. When this happens the water rushes in to fill that 100 mile or so hole which pulls water away from the beach. Next, the rest of the ocean has no problem following the old adage that water seeks its own level. So the hole fills with water, which causes a wave to be created which travels in all directions at speeds up to 600 miles per hour. The series of waves that follow are each less than the preceding wave but still destructive. So if you feel a quake, head for high ground. Or if you see the water runs out quickly, head to higher ground. This is a worst case scenario.

The only tsunami we have experienced in recent years was deceptively small; generated by the 1964 Alaskan quake.

12.10 The Tides

You need to know about tides. Beware the "sneaker waves".

Tides on the Pacific central coast with no factors such as winds or seasonal extremes, move in and out twice a day. The range is high and low enough to make major changes along the coast. The low tide exposes all kinds of secret caves and pathways along the beach. High tide blocks escape routes and can make a nice expanse of sand a dangerous place to be. Living things on the beach and in tide pools are awaiting the next incoming tide so leave them alone! You are so fascinated by what you see all around you that you will forget about tide changes and sneaker waves. Both can move large logs and anything else caught unawares. Better have a tide table with you. This is a weather book, and tides do not cause weather but they can cause you problems.

The gravitational pull of the sun and the moon cause the tides. Despite the mass of the sun, its effect is less than that of the moon as far as tides are concerned. When the moon and sun are on the same side of the earth there exists a higher tide on that side.

There are seasonal changes that move sand out over the winter, exposing cobble beaches, but the sand returns in summer. If you have a favorite rock on the beach you may watch it "grow" as much as head high in winter only to "shrink" to just barely above the sand in summer. Erosion can be a slow process, spanning eons. Along the Oregon coast erosion can strike with the suddenness of a winter storm. Pounding waves and fierce winds carry away thousands of yards of shoreline in a few weeks of time. El Niño years are the times of most erosion. The rugged nature of the Oregon coast prevents the sand from migrating very far along the coast. Seasonally, the sand tends to move offshore and back onshore more than being carried away to other places.

When you come to the coast guess where you are going to end up? On the beach! Remember, the tide has two lows and two highs every 24 hours. This is termed a diurnal tide.

12.11 The Surf

One of the things about the beach that has no obvious relation to the weather is the surf. Don't turn your back on the waves or get on or near a beach log. If you see a six foot diameter log stuck on the beach, it is hard to believe the next wave can move the whole log on top of you. That will be the tragic end to a beautiful day on the beach. What is obvious is the waves on the beach are affected by winds, tides and occasional boat wakes if you are in a bay or river.

As usual, the professional meteorologists have sliced and diced waves so many ways you may not remember all the names. For our use, especially if you listen to the Coast Guard giving reports, some knowledge of wave anatomy is necessary.

The top is the Crest. The hole at the bottom is the Trough. The distance between crests is the Length or Period.

The height of a wave is the vertical distance between the crest and the trough. The distance between successive crests at a stationary point is called wave Period. Waves usually come in groups, called a series by the Coast Guard.

The wind is responsible for altering the water surface. The air moving in contact with the water causes it to pile up in ridges that we call waves. It starts with about a 2 mph breeze and grows into whitecaps at about 15 mph. Swells and waves that are running out of energy do not crest or curl over at the top. For our money swells are the culprits that upset the human digestive system; up and down, up and down. Give us choppy seas and spray flying in the wind. We know the waves can reach fantastic heights. It is possible to have 15 to 20 foot high waves. A moderate storm at sea can send as many as 600 waves ashore in an hour.

One confusing thing is the way a wave moves along. You can see that motion but here is the goofy part. The water molecule goes in a circular motion and, generally speaking, stays in the same place relative to the bottom. Basically, the object goes up and down as the waves pass it. That forward motion we perceive taking us past objects, is caused by the wind.

The bottom depth and the configuration of the beach affect the waves we see. We heard a fisherman talking about the long period in seconds between waves. His suggestion was they must have been generated by the winds a long way out at sea. The day was not especially windy so we feel he probably was correct. The weather guys know storms as far away as Hawaii create waves on our beach.

12.12 Surfer's Wave Lexicon

The surfers have their own nomenclature for waves which center on how well the wave works for them. Their idea is to get as long a ride as possible. If you are near a beach overlook you will see a collection of vehicles. They are of no special make but most have racks on top for surfboards. These folks are usually friendly to others and the environment. They simply enjoy the waves.

Based on some unknown signal they all start showing up and equally amazing, they fade away at various times. Most of the time land lovers can see why the waves are not good, but other times only the surfers know what is good and what is not for their sport.

Surfer talk probably is not known world wide so do not get on our case if in Hawaii they use other jargon. Here they call the perfect wave orgasmic. If the wave breaks too fast to surf it is called a close-out. When the surfers paddle in the wave and come out its called being in the green room. Another term that fits here is getting barrel rolled. When they get hammered and can not get through the wave it is called getting headed. When it is really bad like the inside of the clothes washer and everyone goes home to dream about the perfect wave the phrase is just plain crap.

As you can see Freud, would have fun with their choice of phrases but in today's world it seems pretty innocuous. Our intent is to provide information that helps you enjoy weather fun at the beach. Don't hesitate to talk to the surfers. Some are bald and paunchy which suggests this is not just a kids sport.

Enjoy watching the waves. Stay wary and read more about them. More and more research is going on about what really is happening with 'riptides' and wave phenomena. Rip tides are explained in the Glossary.

12.13 Time

You thought you knew what time it is. You get confused when you cross time zones. Your body tells you one thing and the local clock tells you another. Worst of all, you forget what time it is at home, so you call the kids or spouse while they are sleeping. Then you wonder why they are grumpy. When you watch some TV they show numbers on the corner of the screen titled Zulu time, PST, MT, etc. You are expected to know what that time means.

The United States time zones are Hawaii-Aleutian, Alaska, Pacific, Mountain, Central, Eastern, and Atlantic. These are the TIME ZONES with which we are familiar. Just off the Atlantic shore you have Newfoundland time which is plus one-half hour.

Now let us explore the foundation of time zones. Way back when Britannia ruled the waves, they needed a way to keep track of where they were so they knew they were lord and master. It was relatively easy to figure out the time while you are going south or north. Just measure the angle of some star, usually the North Star. The further South you are the lower the angle. The real problem was how far west you are. The nice lines on the map (longitudes) are not painted on the real earth or the ocean. Someone decided those longitudinal lines would start at zero degrees and all time would be measured from that point. Now we have the Zero degree longitude starting at a place on British soil called Greenwich (pronounced Grenich).

Now the military gets into the act with their phonetic alphabet with 'Zed' or 'Zulu' meaning Zero. Yes, they are combined so Greenwich Time is Zulu and they are both called 'Greenwich mean time'. By the way, we are describing the twenty-four hour clock; not a.m. and p.m.

GREENWICH MEAN TIME (GMT) is the world standard and called the 'Universal Coordinated Time (UTC). Why is it UTC instead of UCT? Remember all this comes from across the ocean where word arrangement differs so Universal Time Coordinated it is, but just remember to say UTC.

When you look at the clock you convert to Zulu or another zone by using the table below. Note, it is one hour increments.

GREENWICH MEAN TIME (GMT) is 0 at midnight. Therefore moving west three zones which are in the Atlantic Ocean,

ATLANTIC	AST is minus four hours so this example is 8 pm
EASTERN	EST is minus five hours so this example is 7 pm
CENTRAL	CST is minus six hours so this example is 6 pm
MOUNTAIN	MST is minus seven hours so this example is 5 pm
PACIFIC	PST is minus eight hours so this example is 4 pm
ALASKA	ALA is minus nine hours so this example is 3 pm

HAWAII-ALEUTIAN; HAST is minus ten hours so this example is 2 pm.

12.14 Weather Data Sources

Want your computer to tell you the weather or to give you abundant data? Well the old www... works here also. Just add Weather Channel, USA TODAY, CNN, NOAA, and others you see in magazines and newspapers. Then you can enjoy trying to 'out guess' the 'experts'.

Here are sources for detailed weather data (historical) and for a range of forecasts.

http//www.

wrh.noaa.gov
nwrfc.noaa
weather.gov
accuweather.com
oregoncoast.com

wunderground.com

Keep in mind there is a limited quantity of independent forecasts. Also these forecasters are, in many cases, using the same data gathering instruments. Remember the sophisticated instruments we talked about in Section 7 herein? One of the things to keep in mind is do you want historical raw data from which you can forecast or do you want to compare others interpretations? We will also provide you with some computer software titles which, in some cases, do both tasks for you. Okay, so now where should you go to on the Internet to get the good information? Keep in mind we are talking about our north central Oregon coast.

Next is the software on your computer. Again, we must remind you that new and old software abounds for computers and the best is the one that does what you want. If you buy a weather station such as a Davis or Oregon Scientific, the manufacturer usually has software packages available de-signed specifically for their instrumentation. So gather the tools and add to your personal folder the various internet sites you like. Then, and we must repeat, then the fun begins because it is you that must make the forecast. That is the real fun of Weather Guessing.

There are weather buoys for us along the northern Oregon coast. One, Buoy #46050, is off Yaquina Bay and the other is near Astoria.

The National Data Buoy Center is an excellent source for current and historical weather data from just off the coast. However, this source has wonderful links to other sources so we consider it our first choice for information and forecast helps here on the Oregon coast.

Always remember that most sites have links to additional sources and sources come and go on the internet. So when you are in the mood, jump on the internet and learn more of where the data is and what it means. Use your search engines to find all the available and plentiful data.

When you ask for specific forecasts we find it is usually from an AccuWeather data base.

12.15 LIGHTNING

Thunder roars. The earth is negatively charged and the air is a poor conductor of electricity. So the charge in the air builds to as high as one million volts per meter (whatever that means) and finally breaks down the air's insulating properties. The resulting lightning stroke is a sudden flow of a negative charge from the cloud to the ground, through a channel of leader strokes. This is followed by a ground to cloud streamer carrying a positive charge and a peak current that may exceed 200,000 amperes. This causes explosive heat expansion of the air and we have the crash and rumble of the thunder. How Far away is the Lightning?

Because of the speed of sound through the air; 1,000 feet per second, the distance in miles to the lightning stroke can be measured.

You count the number of seconds between the sighting of the lightning flash and the sound of the thunder. Divide this number by 5 and there is your answer in miles (approximate).

12.16 Fathoms and Knots

A Fathom is six feet in length and is used for seafarers' measurement of depth of the sea under their vessels. Mark Twain made the riverboat lead line heaver famous by using the phrase mark twain; which means two fathoms or 12 feet. He did not continue on with mark thrice, etc.

Now why did the 'old salts' pick six feet instead of five or ten? The word comes from the old English word of *faedm,* meaning to embrace, and is a measurement across the outstretched arms of a man of average size. The metric system (meters or meters) is rapidly replacing usage of Fathom.

A fathom is equal to 1.8256 meters.

We told you about the lead line, where they count the knots to tell the depth of the water. Well, the same technique was employed to guess the speed of a boat. A chunk of wood was thrown off the stern (boat talk for back end), and they counted how fast the knots in the line paid out as the boat went forward.

Here are some facts about Knot. A Knot's length is 6,080 statute feet

compared to 5,280 feet in a statute mile. For wind speed a Knot is one nautical mile per hour. A Nautical mile is one 60th of a degree of latitude.

KPH = knots per hour

The actual length varies by latitude from 6,046 feet on the equator to 6,092 feet at Latitude 60°. If the world wasn't a globe the sizes would be uniform. Latitude 60° is just about 25 miles north of our location on the North/Central Oregon coast. Yes, that puts Lincoln City, Oregon at Latitude 45°.

The Nautical mile for measuring speed is 6,076.11 feet (1.151 statute miles).

1 KPH = 1.152 MPH = 1 nautical mile per hour

12.17 Sea Stacks

When the ocean wins the battle of erosion the basalt plugs are cut off from the mainland and you get sea stacks or high rocky islands like famous Hay Stack Rock off Cannon Beach. Note that Pacific City also has a fine example of this while the rest of us have rather small off shore rocky islands. Most of our headlands have retained their connection to the mainland.

In Lincoln City we have a rocky island offshore that is probably part of the lava intrusion of Bechia caused during the formation of the volcanic plug now officially named

Roads End point. Further south are more rock islands, but none are as dramatic as the sea stacks that you will see as you travel north on the Oregon Coast.

12.18 Summation

This section is actually a collage of information mostly related to weather. Our purpose is to make you a proficient Weather 'Guesser' for the Oregon coast. To help you do this, we provide some global information that aids in your understanding and researching for local conditions of climate and weather. From Sun to Water to Internet, the information at least allows you to be more knowledgeable. Everyone needs to know about the different time zones. Right? Odds and Ends are the ancillary information.

13. TABLES, GRAPHS, AND CONVERSIONS

Here are some of the particularly significant bits of data for your needs in reading and interpreting the weather data. This is a list and the details follow:

- How to convert temperatures
- What many of the symbols mean
- Different barometer (air Pressure) scales
- Raindrop sizes
- Humidity readings
- Wind speed scales
- Heat Stress charts
- Wind speed conversions and then symbols
- How to read wind speed
- Measuring water content of snow
- How far can you see

13.1. TEMPERATURES

A <u>F</u>ahrenheit degree is <u>smaller than</u> a <u>C</u>elsius (Centigrade) degree.

F<C

One Fahrenheit degree is 5/9 of a Celsius degree.

Fahrenheit to Centigrade - multiply by 0.556 If you wish the quick look approach, it is about half plus a little bit more.

or

subtract 32 from the F°, multiply by 5, and divide by 9.

$((F-32)*5)\div9=°C$

Centigrade to Fahrenheit - multiply the C° by 9, divide by 5, then add 32°

$((C*9)\div5)+32=°F$

13.2 INTERNATIONAL WEATHER SYMBOLS

On weather charts an H marks the center of an area where the atmospheric pressure is higher than its surroundings. Comparatively an L indicates the center of an area where the atmospheric pressure is lower than the surrounding areas. These areas are marked by lines joining places of the same air pressure, called isobars.

Most weather maps note three types of air mass:

1. Maritime Polar is cold air that forms over the ocean which causes coastal fog.
2. Maritime Tropical air is warm tropical air, again formed over the ocean. Maritime Tropical air is a more stable and wet air mass.
3. Continental Polar is cold air formed over land and is the most turbulent.

13.3 BAROMETRIC PRESSURE

The Barometer Is Moving!

Atmospheric pressure changes due to heat variations. Winds move from areas of high pressure to areas of low pressure.

A rising barometer means the air pressure around you is getting heavier so winds diminish and change is less eminent. This is usually a signal for stable and more preferable weather. When the barometer rises quickly we expect clearing. Rain should end BUT humidity can increase so it can become muggy.

Falling air pressure means something is causing the air to move. Higher pressure pushes air toward low pressure. A falling barometer means there is less air pushing down. This signals a change; could be snow, could be rain, or could be wind. Rapidly falling air pressure is a low pressure system so a change in the weather is coming soon.

13.4 RAIN-DRIZZLE-MIST

The water drops in clouds turn into rain because the air is cooled enough to condense the water into liquid. Coalescence forms the bigger drops. Actually, the drops in the clouds grow a hundred times in size by condensation and by the cloud drops absorbing smaller ones as they fall (coalescence).

A rising temperature aloft thins the STRATUS clouds into nimbostratus and we have drizzle instead of rain. This is the lightest form of rain.

Mist is a very low cloud. The mist is slightly finer drops than drizzle so does not fall. Remember this is a cloud and clouds do not fall. Denser mist is Fog.

OK, let's have the drops get big enough to really fall. Gravity wins.

Want to have some fun? Buy a rain gauge and keep track of your own rain amounts. The amount of rainfall varies based on wind eddies, proximity of buildings, trees and the gauge itself. Placement of a rain gauge in an open space is mandatory.

13.5 Humidity

A Hygrometer consists of two thermometers; a dry bulb and a wet bulb. The wet bulb temperature is usually lower because of evaporation from the wet wick around the bulb. If both temperatures are the same, the air is saturated so no evaporation can take place.

Yep! **That** is 100% Humidity.

The greater the temperature difference, the drier the air is, and the greater the evaporation, thus the Lower the Humidity.

Remember <u>the warmer the air the more moisture it can hold.</u>

The Hygrometer Table:

Dry Bulb Temp.	Wet Bulb Temp.	Relative Humidity
85°F	85°F	100%
85	80	78
85	75	47
85	65	33
50	50	
100%	50	45
68	50	40
15	50°F	30°F
1%		

This is a short example of what the Hygrometer Tables contain.

13.6 Feeling the Heat

Many factors contribute to **heat stress**, but the most important elements influencing heat stress and comfort are temperature and humidity. Heat stress charts are available from The Red Cross or similar agencies. These charts use a combination of high temperature and high humidity to estimate the threat of heat stress. Research has lead to development of the Apparent Temperature index which assumes a very light breeze and the person being in the shade. Of course, how hot it feels is a matter of personal judgment, but this index seems to give most people an idea of what the hot weather really feels like.

A typical chart shows both the relative humidity and ambient air temperature. Where the two values intersect you have a value (number) to use as follows:

Greater Than 130 = <u>Extreme Danger</u>; Heatstroke imminent

105-130 = <u>Danger</u>; Sunstroke or heat exhaustion is likely.

With prolonged exposure and physical activity, Heatstroke is possible.

90-105 = <u>Extreme Caution</u>; Sunstroke or heat exhaustion is possible with prolonged exposure and physical activity.

80-90 = <u>Caution</u>; Fatigue possible with prolonged exposure and physical activity.

13.7 Wind Speed Symbols

Weather maps usually have International Weather Symbols.

The wind speed symbol looks like a circle with a tail. The tail length is not significant but it has feathers. The number of feathers tells us the wind speed.

No feathers is 1 to 2 knots of wind

Two feathers is 18-22 knots

Three and a half feathers is 33-37 knots

A triangle on the end of the tail indicates 48-52 knots

13.8 How Wet Is Snow

First let us measure the depth of the snow. This is measured with an ordinary yard stick or ruler at three or more representative areas of snowfall. The recorded snowfall is an average of several measurements. Drifted snow poses special problems and sometimes ten measurements are made to get a representative value. A weighing type of snow gauge is also used. It weighs the snow as it falls and automatically registers the depth according to weight.

Snow can be moisture crammed or very dry. Normally, the water equivalent of snow to water is ten to one. Thus, ten inches of snow would provide a water equivalent of one inch. This ratio can vary considerably. Very wet snow ratio to water is about six to one. Ten inches of extremely fine powdery snow may melt down to about only 1/3 inch of water.

13.9 The Horizon Is How Far Away?

As we all know the old earth is round and curves away from us and our line of sight is straight out from our eye level. The average person's eyes are approximately 5 foot up in the air. At this height you can see approximately 2.9 miles to the horizon. The higher you go the farther you can see to the horizon. Here are some local examples:

- The seaside promenade is about 20 feet up and you can see 6.6 miles.
- On the Depoe Bay bridge you are 58 feet above sea level and can see 10.5 miles.
- On the Yaquina Bay bridge in Newport, Oregon you are 138 feet above sea level and you can see 15.8 miles.
- A little farther south to Cape Perpetua Lookout, you are 800 feet above sea level so your sight distance to the horizon is about 37.3 miles.

Have fun guessing the distance to the horizon when you are watching the weather coming in over the ocean. It is fun to do. Enjoy the view.

QUICK GUIDES TO WEATHER

This is the data you should try to collect:

- Sky conditions
- Barometric pressure
- Temperature
- Wind Speed and Direction
- Wind chill
- Humidity
- Dew point
- Rain totals

These are the steps you should use to gather that data:

1. Look! Do a sky scan & check instruments.
2. Feel! Pay attention to your senses.
3. Observe nature's creatures.
4. Put it all together for an 'educated' guess.

SKY CONDITIONS

LOOK

Big puffy, separated heaps of clouds = cumulus = turbulence.

Growing cumulus = showers later on.

Flat-bottom base clouds = fair weather.

Layered clouds that are wider than thick Stratus = rain may come.

Sky filled with layered, lumpy clouds = Nimbus = rain.

Big puffy clouds with rising peaks = cumulonimbus = unstable

Wispy clouds that are very high = cirrus = a front exists

Cloud names - Remember the pictorial and textural definitions of section 1.4? Having cloud pictures in front of you while looking skyward is always a good idea.

CUMULUS	clearing weather or rain Cumulus clouds go vertical = rain
STRATUS	may thicken into nimbostratus with rain
CIRRUS	behind or leading a front to fair weather
NIMBUS	precipitation
NIMBOSTRATUS	rain or snow
CUMULUSCONGESTUS	rain possibly showers
CUMULONIMBUS	severe weather (maybe tornado)

Next thing, obviously, if there are no clouds in sight, it will be a while before you get rain or simply no clouds, no rain.

If the clouds look fat low and heavy you can bet rain is probable.

FEEL

If you feel the clouds; it is fog and the ground is warmer than the air.

BAROMETRIC PRESSURE
LOOK

Sea gulls hunkered down = low pressure= winds; weather change coming.

Animals in free range seek shelter or are facing away from the incoming storm.

Barometric pressure: steady equals a no change forecast

Hi and steady= no change; LO and steady= a weather front is moving in; FALLING= changing; air will blow in; could be rain; usually a cold front
Rapid changes = fronts moving through
Rising= Warm front. Probably rain.
QUICK RISE = clearing

FEEL

Feeling logy, worn out? Could be <u>high</u> pressure
Lots of air movement? Should mean <u>low</u> pressure
Joints giving you problems? Barometer probably <u>changing</u>

TEMPERATURE

LOOK

Birds with their wings spread out = hot
Fog = cold air above ground
Flowers (and you) wilting = really hot
Rapid changes in the thermometer indicate a front is moving through. Doesn't this sound similar to barometer movement? Yes we are changing air.
Rising temperature is a warm, front and probably rain.
Falling temperature is a cold front so probably no rain but it could snow.

FEEL

Comfortable? Probably 60 to 70°F.
Need a coat? Temperatures in the 40s or below.
Want gloves and a hat too? Probably in the 30s.
Our body's heat sensors are accurate in terms of letting us know what we need to know to stay comfortable. However, they play tricks on us, because they do not tell us what the temperature is, they tell us the difference in

temperature between the last space we were in and the current one.

WIND SPEED AND DIRECTION
LOOK

clouds flashing by = direction change is coming from
light wind = no change or slow change
spin drifts and white caps on the water = change
Keep in mind, the faster the wind blows the quicker the weather will change.
WEST WIND =   change in the weather; possibly clouds coming in
NORTH WIND =   clearing and colder temperatures
WEST NORTHWEST WIND = fair weather
SOUTH WIND INCREASES = rain
SOUTH WIND SHIFTS WEST clearing/warmer
EAST WIND = warmer temperatures but remember, we seldom have a due East wind
EASTERLY TURNING SOUTH = precipitation

FEEL

Temperature shifts also may accompany wind direction shifts;
To the Northwest, colder;
Southeast or Southwest, warmer;
Easterly winds warmer.

Since everything is relative, a northwest wind in summer is not as cold as one in winter. The point to remember is that here on the beach the wind is usually cool because it is usually coming across open ocean before it gets to us.
Breezes = normal air circulation; no weather change.
Wind   = change

WIND CHILL

Cold? Consider wind chill; it is colder for your body than the thermometer indicates. Cover up.

LOOK

Thermometer is low and there is a wind = caution

Birds and animals are sheltering themselves from the wind= caution

High winds and cloud cover = caution

FEEL

When it is icy around you with dew frozen on objects and, if accompanied by wind, this is a danger signal. Perspiration from strenuous activity, with our cool ocean wind, signals wind chill danger. The evaporation acceleration causes the body to loose heat rapidly.

HUMIDITY

LOOK

Lips dry, skin flaky, static electricity = low humidity

Morning fog evaporates = dry, fair weather.

Rapid evaporation is low humidity.

Moisture on the deck railing and windshields (while it is not raining) suggests high humidity.

FEEL

With a higher humidity your temperature sensors give you greater sensation of cold or hot than in a low humidity situation.

Air is moist; it suggests rain, and it should feel warmer.

Skin feels the humidity because evaporation is reduced.

Air is dry, it will feel colder because the moisture on your skin evaporates and lowers body temperature.

Think your hair is thicker or curly hair is frizzy, then there is more water vapor in the air.

Straight hair becomes limp with increases in humidity.

High humidity is heavier air and gives you that logy feeling.

DEW POINT

LOOK

Water vapor condensing on the surface of solid objects indicates the Dew point has been reached.

This indicates higher humidity near the ground. Therefore we should see clear air, day or night.

See frost? = Dew point temperature is at or below freezing.

Heavy frost or dew in early morning = up to 12 hours good weather

Afternoon temperature and Dew point close together = fog that night

The day's Dew point temperature is probably as cold as it will get that night.

FEEL

Uncomfortable? Could be high Dew point temperature (over 70 F°)

Pleasant? Could be Dew point in 50s or lower

RAIN TOTALS

LOOK

Big puddles in yards = lots! i.e. more than 1/2 inch

Ground absorbs it = needed so follows a dry spell

Poor visibility = > 1/2 inch falling

Misty = no threat of accumulation

Big black clouds coming = RAIN coming

High wispy clouds = no rain

Downpours are .30 inches per hour

Moderate is .11 up to .30 inches per hour

Drizzle is .1 inch per hour or less

FEEL

Splatters on your head = heavy drops from very wet clouds

Soaked through your clothes = steady, heavy rain

Tired of it?! = been raining more than a day

SUMMATION

This is a game of estimated guess. Do not let the mixed signals from instruments and observations discourage you. The more you work at forecasting the weather the more accurate your forecasts will be. In fact, you better keep your sense of humor because your favorite TV personality weather guy or gal has the same problem predicting the weather as you do.

This little scenario might take place:

The television meteorologist giving the weekend forecast.

"On Sunday there may be showers, but if the front pushes through early, we might awaken to a gorgeous sunrise."

A reporter called out from the news desk, when will you know for sure what the weather will be like on Sunday?

The weatherman replied, **Monday morning**.

Enjoy the book. Let it be fun.

GLOSSARY

Advection (fog); fog that forms when warm humid air moves inland over cooler surfaces usually in the winter. Advection refers to air moving horizontally. When the dew point is reached, fog forms.

Equinox - <u>autumnal;</u> Start of fall is September 21st. Day and night are almost equal but now the days start getting shorter.

Equinox - <u>vernal;</u> start of spring is March 20 or 21st. Day and night are almost equal but now the days start getting longer.

See also; solstice

Barograph - a recording barometer that uses a stylus to mark on a drum. The stylus line shows the rise and fall of the barometric pressure

Barometer - an instrument used to measure the pressure of the atmosphere. The two main types are the aneroid and the mercurial.

Barometric pressure - the pressure of the air around us which averages 14.7 pounds per square inch.

Beaufort scale - a scale used to classify wind speed. Devised in 1805 by British Admiral Francis Beaufort to classify winds at sea.

Blizzard warning - sustained winds, or gusts of 35 miles per hour or greater, in combination with considerable falling and or blowing snow, for a period of at least three hours. Visibility will often be reduced to less than one-quarter mile. Temperatures under 20° F add a wind chill hazard.

Breezy - a light wind that flutters the flag and then rests.

Cirrus - thin wispy high clouds made up mostly of ice crystals

Climate - an average of weather conditions as they have occurred over decades. Commonly a thirty year average of temperature and precipitation.

Cloud names - *see the chapter in this book on clouds*

Coalescence - the joining together of small droplets of water into larger droplets

Cold front - a weather front in which a colder air mass overtakes and replaces a warm air mass

Crepuscular rays - the scattering effect of particles in the lower atmosphere causing the rays of sunlight to be visible. Usually occurs with a rising or setting sun.

Cumulus - puffy, billowy clouds

Dew point - the air temperature at which the water vapor in the atmosphere becomes liquid

Doppler (effect) - the change in frequency of sound or radio waves when an object is moving toward or away from another object. Weather radar gives imaging from this effect.

Drizzle - droplets of water having a diameter of less than .02 inches, falling close together. They appear to float in the air but, unlike fog, fall to the ground.

El Niño - a warming of equatorial ocean surface that happens around Christmas. It may disrupt weather and cause havoc from droughts to floods, due to the warmer water's effect on the atmosphere.

El Viejo - normal Pacific Ocean
See also el Niño and la Nina

Fog - a cloud that forms at the earth's surface. Water vapor condenses into tiny droplets of water.

Frost - when the air temperature falls below freezing, the moist air will freeze on objects; frozen dew.

Frost bite - freezing of exposed skin, usually accelerated by high wind in air at temperatures below freezing

Gale warning - sustained winds of 34 to 37 knots (39-43 mph)

GMT - Greenwich Mean Time; the local time at Greenwich, England; zero degrees longitude; the prime meridian.

Graupel - small particles of ice that form in clouds or snowflakes. The small particles become rounded through collision with other ice particles. The subsequent buildup of size results in the particles falling toward earth.

Hail - ice particles that form during thunderstorms when

small ice crystals collect. Wind turbulence tosses the particles upward until layers of ice become heavy enough to fall toward earth. *See also sleet*

Halos - *see rainbows*

High pressure = barometric pressure of 30.00 or greater. It usually signals clearer weather is arriving.

Hurricane - a violent storm of tropical origin in which wind speeds exceed 73 mph. In western pacific, hurricanes are called typhoons and in the Indian Ocean they are called cyclones.

Hygrometer - a two thermometer device to measure the humidity by evaporation cooling one thermometer, while the other thermometer is at air temperature. Hygrometer tables convert the temperature differences into a percentage. This is the percentage of water in the atmosphere.

Hypothalamus - a part of the brain that acts like a thermostat.

Inversion - a reversal of normal temperature readings; the temperature increases with elevation rather than decreases. Inversions often lead to smog as the warmer air is above.

Knots - a unit of speed where one nautical mile is 6,076.12 feet an hour. Used in wind speed measurements

La Niña - the cold phase of the El Niño oscillation of ocean temperatures

Lightning - a visible discharge of electricity produced by a thunderstorm

Littoral cell - used to designate stretches of west coast beaches that lie between outcroppings of land effectively segmenting the coastline. These outcroppings stay in place because they are usually volcanic in origin.

Low - barometric pressure usually below 29.00 inches; indicates changing weather, usually storms, approaching

Meteorologist - a person that uses the scientific study of the Earth's atmosphere, especially its patterns of climate and weather, commonly for weather forecasting.

Microclimate - geographical, biological and man made factors often make the climate of a confined space or small geographical area different from the general climate

Micron - a unit of lineal measurement equal to one millionth of a meter

Military time - a 24 hour clock; no am or pm designations

Millibars - a unit of atmospheric pressure equal to one thousandth of a bar (a basic unit of measure for pressure).

Mist - water vapor at or near the surface resembling but not as dense as fog. The droplets are more defined.

Nimbus - clouds named from the Latin for rain bearing

Occluded front is a weather front that is caused by a cold front overtaking a warm front. The warm air is lifted above the earth's surface. When the cold and warm air meets it is a front.

Pogonips - *see table of contents*

Precipitation is moisture in the form of water droplets and ice crystals large enough, and therefore, heavy enough to fall from the clouds to earth. *See also rain.*

Rain - droplets of water that fall from clouds once they are large enough to be affected by gravity

Rainbow - light is refracted by raindrops and then reflected into the air. This bending of the sunlight is most pronounced on shorter wavelengths and our eyes see the blue and violet first. Red is always the outer color of a rainbow. A combination of the reflection and refraction also creates halos and sundogs.

Relative humidity - the ratio between the actual amount of water vapor to the capacity of the air to hold moisture at a given temperature and atmospheric pressure

Riptide - the process where ocean water is pulled or pushed into a funnel shape at shore and the narrow funnel end goes offshore with great pressure. The cause is a combination of wind and tide. The interesting thing is the phenomenon is short lived. If caught in a riptide an experienced swimmer will relax, knowing the strong current will subside and they can again swim to shore. If you must swim do it parallel to the shore until you are out of the rip. Fight a riptide and it wins; it has more endurance.

Satellites (weather) - man made devices that circle the earth recording and sending information related to temperature, pressure, humidity, etc.

Saturation point - the air temperature at which the atmosphere is totally saturated with water vapor
See also dew point

Sleet - precipitation consisting of small pellets of ice with a diameter of less than one quarter of an inch

Snow - solid form of precipitation consisting of ice crystals

Squall - sudden shower or downpour of rain. It can be accompanied with high winds and sometimes snow

Solstice - (summer) - the beginning of summer; June 21st the longest day The earth has its maximum tilt toward the sun.

Solstice - (winter) - the beginning of winter; December 21st; our shortest day. The earth has its maximum tilt away from the sun.

Stationary front - a weather front that is stalled (not moving) due to a lack of pressure differentials

Storm surge - higher than normal tides and wave heights caused by on shore winds pushing the waves toward the beach

Storm warning - issued by the U.S. Coast guard and NOAA when winds above 47 knots (~41 mph) are expected.

Thermograph - a recording thermometer that uses a stylus to record temperature over time on a drum. It provides a visual record for a period of time determined by drum size and rotation speed.

Thermometer - a device for measuring temperature frequently utilizing a column of mercury or alcohol. A dial thermometer usually uses a spring that is sensitive to temperature changes

UTC - coordinated universal time; measured at Greenwich, England and defined as the zero median. It is French in origin.

Virga - falling precipitation (usually ice) that evaporates before reaching the earth. Often wispy strands of water or ice particles from the base of clouds

Vernal equinox - the start of spring when the sun appears to cross the equator; around March 21; days and nights are equal length.

Warm front is a storm front in which the warm air mass overtakes and replaces a cold air mass. *Which is on top? See occluded*

Water year is a concept of measuring rain totals along the west coast that goes from October 1st to September 31st.

Weather - the <u>current</u> conditions of temperature precipitation humidity and wind

Wind chill advisory - warning that wind may cause temperatures of minus 35 to 50 degrees Fahrenheit

Wind chill warning - wind may cause temperatures to reach levels of minus 50 degrees Fahrenheit; life threatening

Winter solstice - December 21; the shortest day; the beginning of winter

Winter storm watch - an alert provided to warn of changing conditions in which travel should be avoided

Zones - climatic areas (e.g. Tropical or coastal) defined by averages of weather peculiar to a restricted geographical area. The common vernacular is rainy, temperate, or dry

BIBLIOGRAPHY

1001 QUESTIONS ANSWERED ABOUT THE WEATHER by Frank H. Forrester. ISBN 0-486-24218-1

A GOLDEN GUIDE, WEATHER by Paul E. Lehr/ R. Will Burnett

ISBN 0-307-24051-7. LOC 61-8327

A FIELD GUIDE TO THE ATMOSPHERE Vincent J. Schaefer Houghton Mifflin ©1981 ISBN 0-395-24080-8

ATMOSPHERE by Vincent J. Schaefer/John A. Day ISBN 0-395-33033-5

ATMOSPHERE, Peterson Field Guides by Vincent J Schaefer & John A. Day

Houghton Mifflin Co. New York 1981 ISBN 0-395-33033-5

AVIATION FUNDAMENTALS Jeppesen Sanderson Inc. ISBN 0-88487-154-1

CLOUDS AND WEATHER, Peterson Field Guides by John A. Day/Vincent J. Schaeffer ISBN 0-395-56268-6

PADI OPEN WATER DIVE MANUAL Professional Association of Diving Instructors

POPULAR SCIENCE August 1997 pp 56-72

RAINS ALL THE TIME by David Laskin. Sasy Watch Books Seattle ISBN 1-57061-063-0 1997

SKYWATCHERS CLOUD CHART. by The Nature Company. no ISBN number; *see For Spacious Skies*

STORM - Irving Krick vs the U.S. Weather Bureaucracy. by Victor Boesen

ISBN 0-399-20636-1

THE BASIC ESSENTIALS OF WEATHER

FORECASTING. by Michael Hodgson ISBN 0-934802-75-0

THE NATURE COMPANY GUIDES WEATHER. by William J. Burroughs et al. ISBN 0-8094-9774-8

THE OLD FARMER'S ALMANAC 1998. ISBN 1-57198-086-5

THE WEATHER BOOK. by Jack Williams. ISBN 0-679-73669-7

THE WEATHER - USBORNE SPOTTER'S GUIDES. ISBN 0-86020-270-4

THE WORLD WE LIVE IN. editor Kenneth MacLeish Andrew Heiskell; Time Inc. 1955

THE VERMONT COUNTRY STORE Catalog

WAVES AND BEACHES The Dynamics of the Ocean Surface. by Willard Bascom. Anchor Books, Doubleday & Co. New York 1964. Library of Congress card #64-11735

WEATHER TALK Naval Meteorology & Oceanography. Stennis Space Center Miss. MS 39529-5005

WEATHER WATCH by Valerie Wyatt. ISBN 0-921103-63-8

WEATHER WATCHER. y Valerie Wyatt ISBN 0-921103-63-8

YOUR WINGS. by Assen Jordanoff. Funk&Wagnells New York and London 1940